Nutrient Reduction and Biomanipulation as Tools to Improve Water Quality: The Lake Ringsjön Story

Developments in Hydrobiology 140

Series editor
H. J. Dumont

Nutrient Reduction and Biomanipulation as Tools to Improve Water Quality: The Lake Ringsjön Story

Edited by

L.-A. Hansson & E. Bergman

Reprinted from Hydrobiologia, volume 404 (1999)

Springer-Science+Business Media, B.V.

Library of Congress Cataloging-in-Publication Data

A C.I.P. Catalogue record for this book is available from the Library of Congress.

ISBN 978-90-481-5313-8 ISBN 978-94-017-2462-3 (eBook)
DOI 10.1007/978-94-017-2462-3

Printed on acid-free paper

This volume is dedicated to
Stellan F. Hamrin for being brave
and stubborn enough to initiate
and carry out the Lake Ringsjön
biomanipulation programme

Hydrobiologia **404:** vii–viii, 1999.
L.-A. Hansson & E. Bergman (eds), Nutrient Reduction and Biomanipulation as Tools to Improve Water Quality: The Lake Ringsjön Story.

Contents

Hydrobiologia **404:** ix, 1999.
L.-A. Hansson & E. Bergman (eds), Nutrient Reduction and Biomanipulation as Tools to Improve Water Quality: The Lake Ringsjön Story.

Preface

Two generations of research staff at the department of Limnology in Lund have been involved in the mission to divert Lake Ringsjön from a state of severe eutrophication with repeated fish-kills, severe algal blooms and in periods, an almost non-existing Secchi depth. The first step, including an extensive program for reduction of external input of nutrients, was initiated at the time when nutrient reduction was assumed to be the only efficient tool for diminishing the effects of eutrophication. The next step, the removal of large quantities of cyprinid fish, was performed at a time when the idea that top consumers affected the eutrophication process through lower trophic levels had won many researchers souls. Hence, 'The Lake Ringsjön Story' is not only a case study, but also a historical document on how contradictory research ideas have competed and finally merged together into the scientific basis we are standing on today. In retrospect, this may now be expressed as: It is not a question of nutrient *or* top consumer regulation, but of *when* and *how much*?

Both the nutrient reduction plan and the biomanipulation (cyprinid fish reduction) were pioneering measures for such a large system as Lake Ringsjön. However, the results have not been thoroughly presented and data, as well as experiences, have been put into long-term storage on bookshelves of consultants and researchers once the national stakeholder reports were written. It was our conviction however, that there was much to learn both by a more thorough data analysis of specific variables and especially by combining variables into a synthesis. Moreover, despite the measures performed in Lake Ringsjön not being completely successful, since the lake still has periods of algal blooms, 'The Lake Ringsjön Story' may provide valuable experiences for future lake rehabilitation programmes. We therefore decided to convince researchers once involved in studies of Lake Ringsjön to transfer their data from the long-term storage to internationally available knowledge. These persuasions were met with enthusiasm among most of the researchers formerly involved in the various phases of the restoration process. We hope that experiences from 'The Lake Ringsjön Story' may serve as a basis for future attempts at improving the water quality of lakes suffering abuse from urban and agricultural activities. Moreover, we strongly believe that proper management of lakes can turn them into important production units, beneficial to society. A first step towards this goal is learning how to best rehabilitate lakes that have served solely as recipients for too long a time.

Acknowledgements

The papers in this volume have all gone through the ususal review process with at least two referees. As editors, we would therefore like to express our sincere thanks to the following persons for their cooperation and valuable comments on one or several manuscripts:

Peter Blomqvist, Uppsala, Sweden; Åge Brabrand, Oslo, Norway; Anna-Kristina Brunberg, Uppsala, Sweden; Stephen R. Carpenter, Madison, Wisconsin, U.S.A.; Larry Greenberg, Karlstad, Sweden; Jukka Horppila, Helsinki, Finland; Mats Jansson, Umeå, Sweden; Erik Jeppesen, Silkeborg, Denmark; Lena Kautsky, Stockholm, Sweden; Per Larsson, Lund, Sweden; Torben Lauridsen, Silkeborg, Denmark; Åke Lindström, Lund, Sweden; Donald J. McQueen, Toronto, Canada; Thomas Mehner, Berlin, Germany; Olof Regnell, Lund, Sweden; Jan Stenson, Gotenburg, Sweden; Michael J. Vanni, Ohio, U.S.A.; Eva Willén, Uppsala, Sweden; Ian Winfield, Ambleside, Cumbria, U.K.

We are also indebted to Marie Svensson for excellent editorial assistance, Larry Greenberg and Birgitta Devlin for checking the language and Christer Nylander for providing the cover illustration. During the editorial process, we were funded by The Fisheries Board of Sweden (to EB) and Swedish Natural Science Research Council (NFR), The Foundation for Strategic Environmental Research (MISTRA) through the VASTRA-programme, and the EU-project "Shallow Wetland Lake Function and Restoration in a Changing European Environment" (SWALE; contract: ENV4-CT97-0420) (to LAH).

LARS-ANDERS HANSSON
EVA BERGMAN
Lund and Karlstad, November 1998

Hydrobiologia **404:** xi–xii, 1999.
L.-A. Hansson & E. Bergman (eds), Nutrient Reduction and Biomanipulation as Tools to Improve Water Quality: The Lake Ringsjön Story.

Introduction

Eva Bergman[1] & Lars-Anders Hansson[2]
[1]*Department of Biology, Karlstad University, S-651 88 Karlstad, Sweden*
[2]*Department of Limnology, Insitute of Ecology, Ecology Building, University of Lund, SE-223 62 Lund, Sweden*

Many lakes have undergone the process of eutrophication in the 20th century, and Lake Ringsjön (55° 52′ N; 13° 32′ O) is no exception, although over the past 30 years several attempts have been made to lessen the eutrophication. Lake Ringsjön is situated at the geographical mid-point of the southernmost county in Sweden, Scania. It has been of interest to limnologists for more than a century, probably making it one of the best studied lakes in Sweden. The first samples from the lake were taken in 1882, which means there is a 115-year long monitoring serie! Although there are interuptions in the series, the data set is without doubt unique. The first reported investigation of Lake Ringsjön was performed by Trybom at the end of the 19th century (1882–1885), in which he focused on phytoplankton and vegetation (Trybom, 1893). This early description of Lake Ringsjön showed that Secchi depths in the summer were generally 1.5–2.5 m, even if occasional algal blooms developed. At that time, there was also dense submerged vegetation in the shallow areas of the lake, and the fish community was dominated by piscivores, even if cyprinids and whitefish were also abundant (Trybom, 1893). It is likely that the eutrophication process was mainly set off as a result of three different changes:

(a) The lowering of the water level in the second half of the 19th century to free more arable land,
(b) The intensified agriculture that developed following the second world war, and
(c) The higher standard of living, including water-closets and the use of phosphorus-based detergents.

During the period 1966–1972, the total-phosphorus level increased from about 50 to above 200 $\mu g\ l^{-1}$, and the summer Secchi depth (water transparency) decreased from about 2 m to less than 1 m.

During the 1970s and early 1980s, different measures were undertaken to reduce the phosphorus loading to the lake. In 1978, chemical treatment of waste-water by plants in the communities of Höör and Hörby was initiated. This was more than 35 years after the decision of the Swedish Government (1941) to improve that waste-water treatment! In 1975, Lake Ringsjön was included in the project 'Sewage-Water Treatment Plant - Recipient' initiated by the Swedish Environmental Protection Agency. The purpose of the project was to investigate how fast and to what extent nutrient enriched lakes improved when waste-water discharge from municipalities decreased due to the installation of chemical and biological treatment. This project was completed in 1978–79. Since no notable improvements in water transparency were observed in Lake Ringsjön over the investigation period 1975-1978, a new project was initiated in 1979, financed by Mellanskånes Planeringskommitté (a regional collaboration between the municipalities in the Lake Ringsjön area). The project involved collaboration between the two Limnological Institutes at the Universities in Uppsala and Lund. Parallel to this project, the Limnological Institute at the University of Uppsala received funding from The Swedish National Science Fundation ('Forskningsrådsnämnden FRN'), to conduct the project Land Use and Water Quality in City Planning ('Markanvändning och vattenkvalitet i samhällsplaneringen'). Somewhat later, in the 1980s, measures were taken to decrease the leakage of nutrients from the farmlands surrounding Lake Ringsjön. Although phosphorus loading decreased from 30 to 10 tons per year and phosphorus concentrations in the lake decreased, there were no effects on water transparency. Consequently, in the second half of the 1980s, discussions on a possible biomanipulation were initiated. In 1987, a 'preinvestigation' began in Lake Ringsjön, and two years later a cyprinid reduction by means of trawling was started in Sätofta Basin. The expectations were that the abundance and size of zooplankon would increase enabling them to more efficiently graze the algae, and thereby make the lake clear.

All remedial measures undertaken have helped in moving the lake away from its former highly eutrophic state, though not as far along as was hoped. Measures expected to restore the lake by decreasing nutrient loading did reduce phosphorus concentration in the water somewhat, but not enough to affect water transparency and produce the clear lake that everybody wanted. Following biomanipulation, nutrient concentrations decreased further and water transparency showed considerable improvement. However, algal blooms still occur in summertime and the high abundances of young fish that have appeared after biomanipulation may eventually force the ecosystem back to where it was before biomanipulation was initiated.

Even though Lake Ringsjön is still not as clear as we would like it to be, we have learned much from the efforts at getting the lake into its present state. In this book, we aim at describing the lake and the different measures performed. We also attempt to explain why the lake reacted as it did to the various measures. We begin with a '*description of the lake*', its catchment area and its history, followed by chapters on '*fluctuations in water and sediment chemistry*'. Thereafter, impressive time series on '*phytoplankton dynamics and waterfowl*' abundances are presented, starting in the 1890s and in 1968, respectively! These descriptive chapters are followed by a chapter on the '*theory behind the biomanipulation programme*'. We then describe the '*planning and execution of the fish removal*', and discuss the impact of this measure on the '*assemblages of fish, zooplankton, benthic invertebrates and macrophytes*', and also the development of '*toxic algal blooms*'. Towards the end of the book, we intend to draw more general conclusions from the specific chapters by developing and applying an '*ecosystem model*' on Lake Ringsjön. Finally, we '*synthesize*' all available knowledge from the lake by presenting our personal views of possible mechanisms involved in the different periods of the Lake Ringsjön story. Although the different chapters benefit from being combined in a book, allowing cross-references and a more holistic perspective of the multitude of aspects on the lake, they are also separate papers in the journal Hydrobiologia. Therefore, some sections, such as the lake description, are similar in each chapter which may seem somewhat repetitive.

References

Trybom, F., 1893. Ringsjön i Malmöhus län dess naturförhållande och fiske (in Swedish). Medd. Kongl. Lantbruksstyr. 4: 1–42.

Hydrobiologia **404:** 1–7, 1999.
L.-A. Hansson & E. Bergman (eds), Nutrient Reduction and Biomanipulation as Tools to Improve Water Quality: The Lake Ringsjön Story.

Lake Ringsjön: its catchment area, its history and its importance

Lars-Anders Hansson[1], Magnus Enell[2] & Eva Bergman[1]
[1]*Department of Limnology, Insitute of Ecology, Ecology Building, University of Lund, SE-223 62 Lund, Sweden*
[2]*ITT Flygt, P.O. Box 1309, SE-171 25 Solna, Sweden*

Key words: Lake Ringsjön, catchment, drainage, lake restoration, nutrient reduction, biomanipulation

Abstract

The aim of this chapter is to present a general description of Lake Ringsjön and its catchment area, serving as a background for the following, more specific chapters. We briefly describe human activities and land use, as well as the human demographics in the catchment area. Moreover, we present important historical events behind the development and present status of Lake Ringsjön. Finally, we describe the importance of the lake as a production unit for 'ecological services', including not only production of fish and drinking water, but also in terms of recreational value, such as swimming, fishing and bird-watching. Although these 'services' are nowadays of less importance in Lake Ringsjön than 40 years ago, due to the eutrophication problems, the lake has the potential to once again become a centre for human activities.

The lake and its catchment area

Lake Ringsjön is situated in southernmost Sweden and consists of three basins, Sätofta Basin, Eastern Basin and Western Basin (Figure 1). Sätofta Basin and Eastern Basin have contact with each other through a shallow sound, and Eastern Basin and Western Basin have contact with each other through a canal (Figure 2). The total lake surface area is about 40 km^2 of which Sätofta Basin represents 11% (4.2 km^2), Eastern Basin 52% (20.5 km^2) and Western Basin 37% (14.8 km^2)(Table 1). Lake Ringsjön has 14 tributaries, of which five have Sätofta Basin as a recipient, seven discharge into Eastern Basin and two into Western Basin. The lake is drained through the river Rönne å, which runs westwards from Western Basin and empties into in the sea (Kattegatt), near the city of Ängelholm.

The normal water level of the lake is about 54 m above sea level. As is the case with most other lakes in the agricultural areas of southern Sweden, Lake Ringsjön is shallow, with a mean depth of only 4.7 m (Figure 2). The deepest basin is Eastern Basin with a mean depth of 6.1 m and a maximum depth of 16.4 m. Sätofta Basin and Western Basin have mean depths of 3.0 and 3.1 m, and maximum depths of 17.5 m and 5.4 m, respectively (Table 1). Stratification occurs in Sätofta Basin and Eastern Basin, but rarely in Western Basin. The volume of Eastern Basin constitutes 68% (124.8 Mm^3) of the total volume, whereas Sätofta Basin and Western Basin constitute 7% (12.8 Mm^3) and 25% (46.6 Mm^3), respectively (Table 1). The shallowness of the lake is also illustrated by the fact that only 27% and 23% of the volumes in Sätofta Basin and Eastern Basin, respectively, are deeper than the mean depth. The small areas of deep waters in Sätofta and Eastern Basins indicate that lake bottoms dominated by erosion and transport processes are large in these basins. Håkansson (1981) estimated erosion bottom areas to be 30% in both Sätofta Basin and Eastern Basin, and 35–40% in Western Basin. The small difference between the mean depth (3.1 m) and maximum depth (5.4 m) in Western Basin reflects the lack of any specific deep areas, which can be found in the other basins. The retention time is also rather short, varying between 0.2 and 0.5 years in Sätofta Basin and Western Basin, and 0.5 and 1.2 years in Eastern Basin (Enell & Fejes, 1996). The water discharge in the tributaries to Lake Ringsjön varied between 90 and 240 × 10^6 m^3 yr^{-1} during the period 1976 and 1996, and in the outlet the mean water discharge varied between 60 (minimum 19, maximum 117) and 151 (minimum 25, maximum 660) × 10^6 m^3 yr^{-1} during the same period (Enell & Fejes, 1996). There is a low between-year-

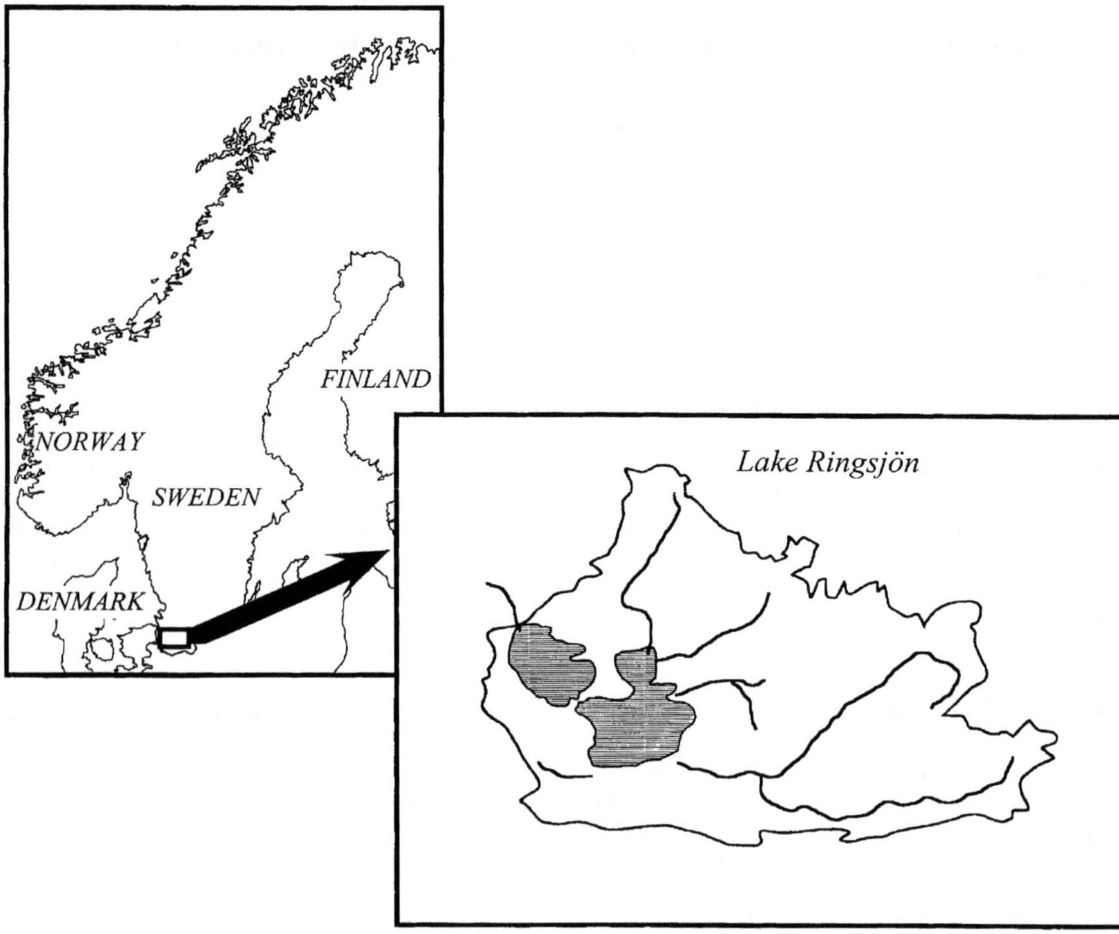

Figure 1. The location of Lake Ringsjön in southern Sweden and a detailed map of the catchment area of Lake Ringsjön, and some of the larger streams and rivers entering and leaving the lake.

variation in the minimum outlet water discharge as the lake is regulated.

Geology and climate

Lake Ringsjön is situated at the fault between the Scandinavian primary rock area in the northeast and an area with sedimentary rock in the southwest, created about 240 million years ago (Triassic). The earth layer is generally thin (less than 10 m), but locally the thickness may reach 40–50 m. The north and eastern parts of the catchment area are covered with nutrient-poor fine moraine sand. In the middle of the catchment area, a band running from north to south of nutrient-rich clayed moraine has created a fertile agricultural area. In the eastern parts of the catchment area, peat bogs are common.

The climate in southern Sweden is characterized by humid, windy and cool winters, and warm and humid summers. The mean annual temperature for the area, calculated for the period 1931–1960 and 1975–1980, was 7.0 °C. The mean annual precipitation for the period 1931–1969 was 689 mm, and for the period 1975–1980 was 710 mm.

Land use

The catchment area of the lake is 348 km^2, of which forests and arable land constitute about 40% each (Table 1). The northeastern part of the catchment area

Table 1. Characteristics of the Lake Ringsjön drainage area. Based on data from Ringsjökommittén (1991). Turnover times are presented as minimum and maximum values during the period 1984–1996

	Sätofta	Eastern	Western	Total
Surface area (km^2)	4.2	20.5	14.8	39.6
Catchment area (km^2)	101.7	221.1	24.9	347.7
Forests	56.5	69.8	8.7	134.8
Agriculture	19.3	100.0	11.1	130.3
Wetland	2.3	6.9	0	9.2
Lakes	1.0	0	0	1.0
Urban areas	2.8	1.4	0	4.2
Miscellaneous	17.7	44.4	5.1	67.2
Shoreline (km)	7.8	21.6	19.1	48.5
Maximum depth (m)	17.5	16.4	5.4	17.5
Mean depth (m)	3.0	6.1	3.1	4.7
Volume (10^6 m^3)	12.8	124.8	46.6	184.2
Turnover time (yr)	0.18–0.46	0.51–1.19	0.18–0.39	0.66–1.49

is at a higher altitude (between 110 and 150 m above see level) than the southwestern part (between 60 and 100 m above see level). These differences in altitude result in the climate being cooler and more humid in the northeastern parts (Nihlgård & Martinsson, 1976; Ryding, 1983).

Land use is dominated by agriculture in the southern part and forestry in the northern part of the catchment area. Agriculture has, during the last century, been the most important activity in the area, due to the fertile soil in combination with a suitable climate. The lake is located in a naturally nutrient-rich soil area, and therefore agricultural activities have been going on for centuries. The development of agriculture has gone through comprehensive changes during the last 50 years, with regard to the use of commercial fertilizers, heavy farm machines and the drive towards ever-increasing harvests per unit area. Since the late 1970s, agriculture's contribution to nutrient inputs in the drainage area has decreased. Commercial fertilizers are used less, and application methods of fertilizers and manure have improved so that leakage to the surroundings is minimized. Much of the farmlands are 'green' during larger parts of the year, and ditches between the farmlands have been improved (Ringsjökommittén, 1991).

In contrast to agriculture, the industrial activities in the catchment area have always been limited and therefore of minor importance for the deterioration of the water quality in Lake Ringsjön. All industries in the area are connected to the municipal sewage treatment plants.

Population in the catchment area

In 1990, the total human population living in the catchment area was estimated to be about 21 000. About 15 000 of these lived in urban areas and their homes were connected to the municipal sewage water systems in the cities of Höör and Hörby. The remaining people (6000 persons) lived in rural areas. A study in 1986 (Länsstyrelsen i M-län, 1987) showed that about 26% of the households in rural areas did not fulfill the standards for water and sewage water management. About 4500 people live in 2200 summer cottages, located mainly along the northern parts of the lake. These summer cottages are in use approximately 4–5 months per year.

A comprehensive program was conducted to connect the homes in rural areas to the municipal sewage treatment plants in Eslöv, Höör and Hörby. A total of about 640 households were connected during 1985–1993, which corresponds to 2000–2500 people. In addition, about 50% of the households that were not connected to the municipal sewage treatment plants were given recommendations on how to improve their waste water treatment, and by January 1989 most of these households had complied with these recommendations. As a result, the input of phosphorus and nitrogen into the drainage area has been reduced by

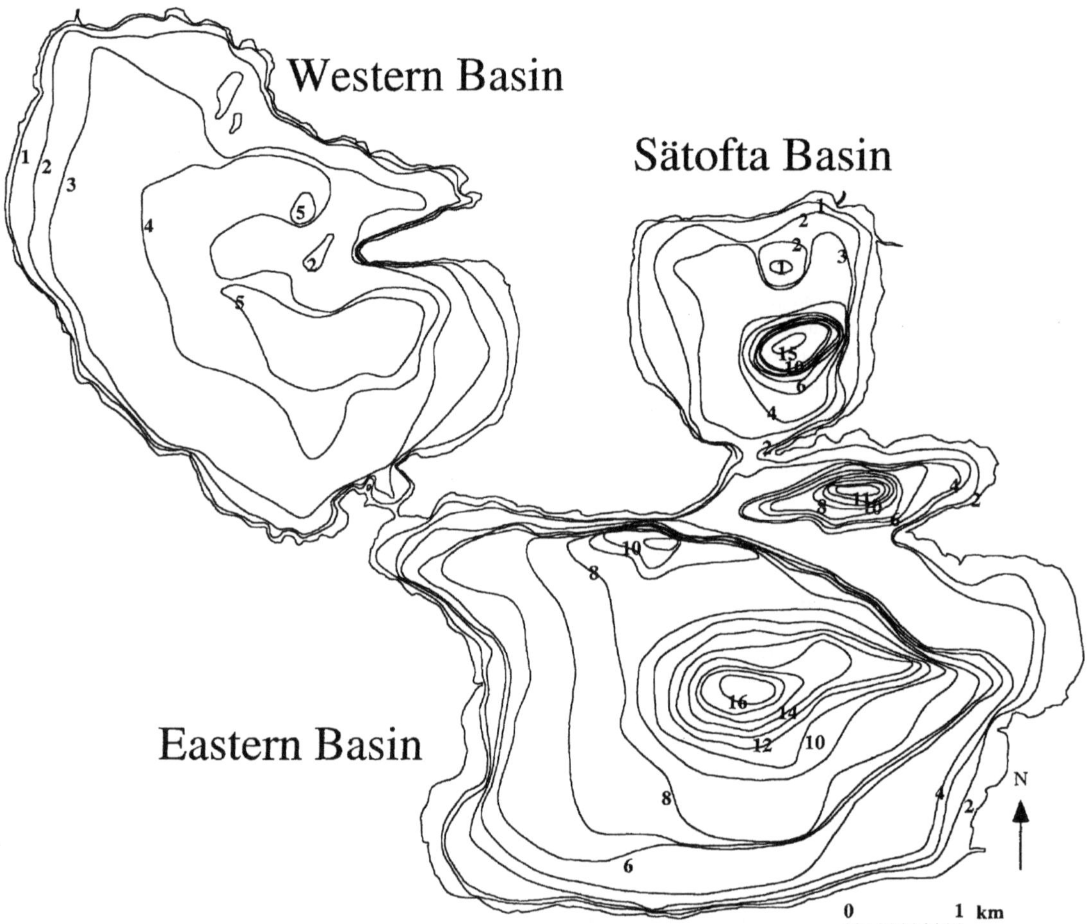

Figure 2. The three basins of Lake Ringsjön: Sätofta Basin, Eastern Basin and Western Basin. Depth contours for each meter are given.

2.5 tons and 4 tons, respectively (Ringsjökommittén, 1991).

Reasons for the eutrophication process

Explanations for the causes of eutrophication of Lake Ringsjön have for the last three decades focused on nutrient leaching because of agricultural practicis and discharges from insufficiently treated water from municipal sewage treatment plants (Enell & Fejes, 1996). An additional factor is the lowered water level caused by water extractions for drinking water and release of water downstream of Lake Ringsjön to guarantee a minimum water flow in the river Rönne å (Enell & Fejes, 1999; see also Bergman, 1999). Moreover, the lake has also received sewage water from households in rural areas as well as summer houses.

History

The historical record of Lake Ringsjön is long (Trybom, 1893), and some important events are summarized in Table 2. In the beginning of the 1880s, the lake water level was lowered by 1.5 m to obtain more arable land (Table 2). This water lowering corresponded to a water volume of about 25%. This, together with the implementation of new agricultural methods during the 1940s, led to an eventual increase in the productivity of the lake – thus, the eutrophication process started. Around 1900, the Secchi depth

Table 2. Brief history of Lake Ringsjön. Information based on Enell and Fejes (1999 and references therein), and Enell unpubl

1883	The water level was lowered by 1.5 m.
1892	Trybom (1893) described the algal bloom in terms that could characterize the lake as mesotrophic-eutrophic.
End of 1890	The transparency was 2–2.5 m during summer. Occasional blooms of blue-green algae occurred. Blooms of diatoms were common. Submerged vegetation was common. The fish community was dominated by perch and pike. The fish catch was about 38 tonnes yr^{-1}, mainly eel, pike, bream, whitefish, roach. Crayfish was also caught.
1914–1923	The fish catch decreased from 57 to 22 tonnes yr^{-1}. The reason for this decrease is not known, but could be over-fishing.
The 1940s	Water transparency during summer was 1.5–2.5 m. Submerged vegetation was still common, but had decreased in quantity and areal distribution. The phytoplankton community showed a tendency to change towards blue-green algae.
End of 1950s	The crayfish plague eliminated the crayfish from the Western part of the lake, where it used to be most abundant.
1960s	The phytoplankton community was changing drastically: Chrysophycées disappeared and blue-greens increased in number and quantity. The phosphorus concentration (annual mean) in the surface water was 25–50 $\mu g\ l^{-1}$. The external phosphorus load was estimated to be at a minimum of 17 tonnes yr^{-1}. The quantity of cyprinids taken away by the fishermen increased from 10–20 tonnes yr^{-1} to around 30 tonnes yr^{-1}.
End of 1960s	Water transparency during summer decreased drastically to about 1 m. Blue-green algae appeared in heavy blooms. Cattle drinking the lake water got ill, and many died due to toxins from the blue-green algae.
The 1970s	Water transparency during summer decreased to 0.5 m. Heavy blooms of blue-green algae during the summer, and the summer phosphorus concentration in the surface water was 200–400 $\mu g\ l^{-1}$. The external phosphorus load was 25–30 tonnes yr^{-1}. The fish catch was 55–80 tonnes yr^{-1}, of which about 50% were cyprinids. The number of perch first decreased and then increased, and the size of perch and roach decreased. Benthic invertebrates decreased in abundance and diversity.
1975	The sewage water treatment plant for the municipality of Hörby was put into operation (chemical precipitation with post-filtration).
1978	The sewage water treatment plant for the municipality of Höör was put into operation (chemical precipitation with post-filtration).
1980	The 'Ringsjö committee' was formed.
The 1980s	Water transparency during summer was around 0.5 m, as a consequence of heavy blue-green algal blooms. The summer phosphorus concentration was 150–250 $\mu g\ l^{-1}$, and the external phosphorus load was 10–15 tonnes yr^{-1}. The fish community was dominated by small-sized populations of roach and bream, and the abundance of carnivorous fishes was low. The fish catch decreased to 20–40 tonnes yr^{-1}, and the catches of perch decreased and that of pikeperch increased.
1985	'Lex Ringsjön' was formulated.
1988	A fish kill occurred in the Eastern Basin during winter/spring 1988 (about 500 tonnes).
1989–1990	A cyprinid fish removal (biomanipulation) was performed in Sätofta Basin from September 1989 to July 1990 (about 100 tonnes).
1992	A cyprinid fish removal (biomanipulation) was performed in Western Basin (about 120 tonnes).
1990-1996	Water transparency during summer increased to 0.8–2.1 m, and the quantity of blue-green algae decreased. The summer phosphorus concentration was 50–100 μg^{-1}, and the external phosphorus load was 5–10 tonnes yr^{-1}. The fish catch increased, a major part (30–50%) of the catch were cyprinids. The catch of pikeperch (piscivorous fish) increased. Modest increase in establishment of submerged vegetation, on a local scale, was observed.

of the lake was still between 2.0 and 2.5 m during summer, although occasional algal blooms occurred (Andersson, 1948; Table 2). Submersed macrophytes formed dense stands, especially in Western Basin and in Sätofta Basin, and the fish community was dominated by perch (*Perca fluviatilis*), pike (*Esox esox*) and roach (*Rutilus rutilus*). Bream (*Abramis brama*) and whitefish (*Coregonus oxyrhynchus*) were also common. During the period 1966–1972, the turbidity of the water in Lake Ringsjön increased considerably, and in 1975 the phosphorus concentration was five times higher than ten years earlier (Bergman, 1999). In the middle of the 1960s, the hygienic conditions of the water in Lake Ringsjön were reported as unsuitable for recreational swimming. At the end of the same decade (1968), cattle that had been drinking water from the Sätofta Basin showed signs of being poisoned (Ryding, 1983). Hence, as for many other lakes in western Europe and North America, the late

1960s and early 1970s was the period when the rate of deteriorating water quality was at its worst.

Due to its severe problems with eutrophication, Lake Ringsjön has been a focus of concern from local, regional and national authorities (Ryding, 1983). As in most other urban areas in Sweden, sewage treatment plants were built in the 1960s and 1970s, but not until 1975 and 1978, respectively, was the chemical treatment of sewage water introduced in Hörby and Höör. In 1980, local and regional authorities, together with the main stakeholders of the catchment area formed the 'Ringsjö committee'. The main aim of the committee was to initiate and coordinate measures toward improved water quality in the lake (Ringsjökommittén, 1991). One of the more important measures in this direction was the formulation of 'Lex Ringsjön', which was introduced in 1985. This 'law' stated that Lake Ringsjön and its catchment area should be regarded as an "especially pollution sensitive area". The goal with this 'law' was to systematically reduce nutrient leakage from point sources, such as tributaries and farms, and to improve agricultural methods and thereby reduce the external input of phosphorus and nitrogen to the lake (Ringsjökommittén, 1991). The external phosphorus load, as well as the phosphorus concentration in the lake, decreased considerably during the 1980s (Enell & Fejes, 1999). Despite this, few variables of water quality showed a positive response, and if they did, it was quantitatively small (e.g. phytoplankton) (Cronberg, 1999; Bergman, 1999; Bergman & Bergstrand, 1999). In order to increase the rate of improvement in water quality, a cyprinid reduction program was suggested in the late 1980s. It should be noted that at this time, the debate was intense as to whether or not biomanipulative tools were efficient in lake restoration. In addition, it was argued that such a large lake as Lake Ringsjön was not a suitable subject for the testing of predictions from scientific theories (Bergman et al., 1999). Despite this, a cyprinid fish reduction program was performed from 1988–1992. Sätofta Basin underwent large-scale fish removal during 1989–1990, and a similar removal was performed in Western Basin in 1992. A major fish-kill removed fish from Eastern Basin in the spring of 1988 (Hamrin, 1999).

Importance

Lake Ringsjön is not only the geographical center of Sweden's southernmost county – Scania, it has also served as an important *recreational area* of great local and regional interest. Recreation activities in the lake include swimming, boating and fishing, which were all threatened by the accelerating eutrophication. *Commercial fishing* has a long tradition in the lake (Hamrin, 1999), and fresh fish, as well as refined fish products, are sold in a store close to the channel between Eastern Basin and Western Basin. In comparison with most other Swedish lakes, fish production is high in Lake Ringsjön and National Fisheries Board gave it status as a "nationally interesting lake", due to its high fish production (Ryding, 1983; Ringsjökommittén, 1991). *Recreational fishing* was very important during the 1960s, and up to 4 000 fishing licences per year were sold (Ringsjökommittén, 1991). After water quality deteriorated, recreational fishing dropped markedly as few fishing licences were sold during the period 1973–1989. However, during the 1990s, the interest for recreational fishing has once again increased.

Due to its geographical position close to many big cities (Malmö, Helsingborg, Kristianstad and Lund, with altogether more than 1 million inhabitants), and its beautiful surroundings, the lake has been an appreciated *swimming lake*. The eutrophication process has, however, reduced the interest for swimming in the lake, although no general swimming restrictions have been given during the last decade. Before 1970, Lake Ringsjön was important as a resting lake for large numbers of *waterfowl* (Andersson & Nilsson, 1999). As with all other recreational activities centred on the lake, bird watching lost its appeal from about 1970.

From 1963 to 1987, Lake Ringsjön was used as a *drinking water source* for several large cities in Scania. Already in 1955, the three cities of Helsingborg, Landskrona and Eslöv asked for permission to extract potable water from the lake. Permission was granted in 1957, and 500 l s^{-1} were allowed to be extracted from the lake during the period 1960–1979, and up to 660 l s^{-1} from 1979 and onwards. Withdrawal of water started in 1963. An additional request for extraction of up to 1 125 l s^{-1} was submitted in 1961, approved in 1972, and valid from 1974 and onwards (or "forever", as is written in the permit) (Ringsjökommittén, 1991). The number of people using the Lake Ringsjön water supply increased successively during the period from 1963 onwards. During the early 1970s, the water treatment plant had to be renovated to deal with the deteriorating water quality in Lake Ringsjön. Chemical treatment and fast filtration were now used instead of the slow filtration technique used before.

In the beginning of the 1980s, about 250 000 persons used the water supply, which meant that a volume of 25–30 Mm^3 of water was withdrawn annually. In October 1987, drinking water was obtained from Lake Bolmen (situated about 200 km north of the Ringsjön area) instead of Lake Ringsjön. Since 1987, Lake Ringsjön has mainly been used as a reserve drinking water source, which is a direct result of its bad water quality.

In summary, the recreational as well as commercial value of the lake decreased in importance as a result of the eutrophication process. However, the measures that have been undertaken have started to once again increase the value of the lake. It is our belief that more measures can be undertaken to once again make Lake Ringsjön a mid-point and a centre for various kinds of human activities.

References

Andersson, A., 1948. Näringstillgång och planktonutveckling i några skånska sjöar. Vattenhygien 4:9-21.

Andersson, G. & L. Nilsson, 1999. Autumn waterfowl abundance in lake Ringsjön 1968–1996. Hydrobiologia 404: 41–51.

Bergman, E., 1999. Changes in nutrient load and lake water chemistry in Lake Ringsjön, Southern Sweden, during 1965 to 1994. Hydrobiologia 404: 9–18.

Bergman, E. & E. Bergstrand, 1999. Lack of a top-down effect on the zooplankton community after a cyprinid reduction. Hydrobiologia 404: 77–87.

Bergman, E., L.-A. Hansson & G. Andersson, 1999. Biomanipulation in a theoretical and historical perspective. Hydrobiologia 404: 53–58.

Cronberg, G. & H. Annadotter, L. Lawton, 1999. The occurrence of toxic blue-green algae in Lake Ringsjön, south Sweden. Hydrobiologia 404: 123–129.

Cronberg, G., 1999. Qualitative and quantitative investigations of phytoplankton in Lake Ringsjön, Scania, Sweden. Hydrobiologia 404: 27–40.

Enell, M., 1991. Ringsjön - Ekosystem i förändring 1990. Beskrivning av sjöns struktur och funktion 1990 samt förändringar under 1975–1990. Mimeographed. IVL AB. (in Swedish).

Enell, M. & J. Fejes, 1996. Ringsjön - ekosystem i förändring. Analys av tänkbara orsaker till sjöns eutrofiering. Mimeographed Streckodsspecialisten i Sverige AB, IVL AB (in Swedish).

Enell, M. & J. Fejes, 1999. Exemplet Ringsjön - vad har vi lärt oss? Effekter i vatten. Mimeographed. Kungliga Skogs- och Lantbruksakademien 138(2): 85–97 Stockholm (in Swedish).

Hamrin, S., 1999. Planning and execution of the fish reduction in Lake Ringsjön. Hydrobiologia 404: 59–63.

Håkansson, L., 1981. Sjösedimenten i recipientkontrollen. Principer, processer och praktiska exempel. Statens Naturvårdsverk, Rapport. PM 1398.

Länsstyrelsen i M-län, 1987. Redovisning av genomförda vattenvårdsåtgärder i Ringsjöområdet. Dnr 11.189-33-87.

Nihlgård, B. & I. Martinsson, 1976. Från snyltrot till enefälad. Axplock ur Ringsjöbygdens flora och vegetation. - Skånes Natur, årsbok (in Swedish).

Ringsjökommittén, 1991. Ringsjöns restaurering 1980–1990 (in Swedish).

Ryding, S-O., 1983. Ringsjöområdet - ekosystem i förändring. Redovisning och utvärdering av limnologiska undersökningar. Mellanskånes Planeringkommittee och Limnologiska institutionen, Uppsala Universitet. Januari 1983 (in Swedish).

Trybom, F., 1893. Ringsjön i Malmöhus län dess naturförhållande och fiske (in Swedish). Medd. Kongl. Lantbruksstyr. 4: 1-42.

Hydrobiologia **404:** 9–18, 1999.
L.-A. Hansson & E. Bergman (eds), Nutrient Reduction and Biomanipulation as Tools to Improve Water Quality: The Lake Ringsjön Story.

Changes in nutrient load and lake water chemistry in Lake Ringsjön, southern Sweden, from 1966 to 1996

Eva Bergman
Department of Limnology, Insitute of Ecology, Ecology Building, University of Lund, SE-223 62 Lund, Sweden. Present address: Department of Biology, Karlstad University, S-65188 Karlstad, Sweden

Key words: P-loading, P concentration, sewage water treatment plant, agricultural fertilization control program, fish reduction, Secchi depth

Abstract

Monitoring Lake Ringsjön's water chemistry, i.e. of pH, PO_4-phosphorus, total phosphorus, Kjeldahl nitrogen, NO_2+NO_3-nitrogen, total nitrogen, chlorophyll *a* concentration and Secchi depth, has been conducted on a relatively continuous basis since the mid 1960s. Here, I analyze this data by graphical methods and discriminant analysis. The results show that there was a pronounced decrease in Secchi depth and an increase in phosphorus and chlorophyll *a* concentrations during the late 1960s and early 1970s. To improve conditions in the lake, new sewage water treatment techniques, using chemical precipitation techniques with post-filtration, were put into operation in the late 1970s in Hörby and Höör. A few years later, new recommendations were given to the farmers about use of fertilizers to further decrease the nutrient load to Lake Ringsjön. The concentrations of phosphorus and chlorophyll *a* started to decrease in the early 1980s, but there was no increase in Secchi depth. From 1989 to 1991, Secchi depth increased in the three basins of Lake Ringsjön. This increase in Secchi depth co-occurred in time with: (a) a fish-kill in Eastern Basin (1988), fish reductions in Sätofta Basin (1990) and Western Basin (1992); (b) a very low phosphorus loading in 1989 (4.7 tons); and (c) phosphorus concentrations in the lake water below $100\ \mu g\ l^{-1}$.

Introduction

Lake Ringsjön has been of interest for limnological studies for more than a century. The first reported investigation was performed by Trybom between 1882 and 1885, and dealt with phytoplankton and vegetation (Trybom, 1893). Additional phytoplankton studies were done during the 1920s by professor E. Naumann, and in 1946–1947, studies on macrophytes, phytoplankton and hydrochemistry were performed (Andersson, 1948; Almestrand & Lundh, 1951; Lundh, 1951a,b). In the 1960s, physical and chemical variables were also included in the investigations (Björk & Lettervall, 1968), and from the early 1970s several different investigations were conducted. In 1975, Lake Ringsjön was included in the Swedish Environmental Protection Agency's project 'Sewage Water Treatment Plant–Recipient' ('RR-investigation') (Ryding, 1983). The purpose of this project was to investigate how fast, and to what extent, nutrient enriched lakes improved following a decrease in the external nutrient load from municipalities, due to installation of chemical and biological treatment at sewage water treatment plants. In 1983, the control program was modified, and was conducted by The Department of Limnology, University of Lund in 1983–84 (Enell, 1985). From 1985 to 1996, the same program was conducted by the Swedish Environmental Research Institute in Stockholm (Enell, 1986; Enell & Eriksson, 1987, 1988; Enell & Henriksson, 1989, 1990, 1991, 1992; Enell et al., 1993, 1994; Enell & Fejes, 1996). In 1987, a biomanipulation project started, conducted by The Department of Limnology, University of Lund (Hamrin et al., 1991, 1993; Bergman et al., 1994; Bergman, 1997).

In this paper, I focus on the period 1966–1996 because data on physical and chemical variables are available, and because the lake became eutrophicated during this period, and efforts were made to improve lake water quality. In the 1970s, sewage water treatment plants in Hörby and Höör were renovated to include chemical precipitation with post-filtration. Subsequently, new regulations about the usage of fertilizers for farmers working in the drainage area were introduced. A fish-kill occurred in Eastern Basin in 1988, and cyprinid fish were selectively removed from the other basins between 1989 and 1992. In this paper, I present changes in both loading and nutrient concentrations from 1966 to 1996, and I use discriminant analysis to identify any differences in water quality corresponding to the different remedial measures.

Study site

Lake Ringsjön is situated in southernmost Sweden (55° 52′ ; 13° 32′) and consists of three basins: Sätofta Basin (4.2 km^2, mean depth 3.0 m, maximum depth 17.0 m), Eastern Basin (20.5 km^2, mean depth 6.1 m, maximum depth 16.4 m) and Western Basin (14.8 km^2, mean depth 3.1 m, maximum depth 5.4 m) (Hansson et al., 1999). As is the case with many other lakes in urban and agricultural areas, Lake Ringsjön became eutrophicated during the 1960s and 1970s. In the beginning of the 1980s, a *nutrient reduction programme* was initiated which reduced the external input of nutrients to the lake from phosphorus loadings exceeding 30 tons per year to around 10 tons per year. About ten years later (1989–1990), a *cyprinid fish reduction programme* (biomanipulation) was carried out in Sätofta Basin removing about 100 tons of fish (about 60% of the standing stock). During the winter of 1988, there was a fish-kill in Eastern Basin, eliminating about 500 tons of fish (about 80%) (Hamrin, 1999). The biomanipulation in Western Basin, which was performed in 1992, removed about 120 tons of fish (about 50%).

Materials and methods

Water samples were taken for analyses of pH, PO_4-phosphorus, total phosphorus, Kjeldahl nitrogen, NO_2+NO_3-nitrogen, total nitrogen and chlorophyll *a* concentration. In addition, Secchi depth was measured on each sampling occasion. In 1966–1967, samples were taken at a depth of 1 m above the deepest part of each of the three basins using a 5 l Rodhe-sampler at a depth of 1 m, and the analyses were done at the Department of Limnology in Lund, Sweden. During this, approximately one-year period, 12–13 samples were collected, somewhat more were taken in the summer, when samples were taken every three weeks (Björk & Lettervall, 1968). Data from 1975 to 1980 were collected for the 'RR-investigation' (sewage treatment works - recipient - investigation) (Ryding, 1983) and were taken once a week from May to October every year. Samples were taken at a depth of 0–2 m in the central parts of the basins. These water samples were analyzed at the Department of Physiological Botany in Uppsala, Sweden (Ryding, 1978, 1983). During the period 1981–1996, water samples were taken every month, except between 1987 and 1994 when samples were taken twice a month (Enell, 1985, 1986; Enell & Eriksson, 1987, 1988; Enell & Henriksson, 1989, 1990, 1991, 1992; Hamrin et al., 1991; Enell et al., 1993, 1994; Hamrin et al., 1993; Bergman et al., 1994; Enell & Fejes, 1996; Bergman, 1997). Water samples were collected during this period from the uppermost 2 m above the deepest parts of the three basins. All these samples were analyzed at the Department of Limnology in Lund, Sweden. In all cases, that is from 1966 to 1996, the analyses were performed in accordance with SIS-methodology.

Water samples for analyses of nutrient loading of Lake Ringsjön were taken weekly in all larger streams that flow into Lake Ringsjön. Water level was recorded in each stream at each sampling occasion. Water discharge in Hörbyån was recorded continuously by SMHI (Swedish Meteorological and Hydrological Institute), and in the other streams, water discharge was estimated from discharge - stream area curves for streams of known discharge (Ryding, 1983). The amount of water removed for consumption by man was calculated using data during the time the pumps in the water plant were working. Data on precipitation, temperature and wind were provided by SMHI, and evaporation was calculated in accordance with Thornwaite (1948). Estimates of nutrient loading started in 1976 during the 'RR-investigation' (Ryding, 1983), and continued as part of the monitoring program (Enell, 1985, 1986; Enell & Eriksson, 1987, 1988; Enell & Henriksson, 1989, 1990, 1991, 1992; Enell et al., 1993, 1994; Enell & Fejes, 1996).

Changes in the lake's water quality were analysed using graphical methods based on summer means (=May–September) during the period 1966–1996. In addition, I analyzed the data using discriminant analysis (DA) in order to determine if the different periods in which various measures were implemented in Lake Ringsjön could be identified by corresponding changes in water quality.

Table 1. The different periods used in the discriminant analyses. Note that Sätofta Basin only has four periods. Period *a* represents the period before eutrophication begun, period *b* is the period after eutrophication begun, but before anything was done to lower phosphorus loading, period *c* represents the period after the sewage water treatment plants in Hörby and Höör included chemical precipitation with post-filtration, period *d* represents the period after regulations about the usage of fertilisers for farmers in the drainage area were introduced, and period *e* represents the period after the fish reductions (the year when the fish reductions occurred are omitted from the analyses)

Periods	a	b	c	d	e
Sätofta Basin	1966–67	1973–82		1983–89	1991–96
Eastern Basin	1966–67	1973–75	1976–82	1983–87	1989–96
Western Basin	1966–67	1973–77	1978–82	1983–91	1993–96

The discriminant analysis was based on a 30 year period, i.e. 1966–1996. In this analysis, I analyzed Secchi depth, PO_4-phosphorus, total phosphorus, Kjeldahl nitrogen, NO_3+NO_2 nitrogen and pH. Chlorophyll *a* was excluded as data was missing for too many years. Each basin was analyzed separately since the various nutrient reduction measures were performed at different times for the three basins. The DA was based on four periods for Sätofta Basin and five periods for Eastern and Western Basins (Table 1). The first period is before the lake became eutrophicated (period *a*). The second period (*b*) began when nutrient levels increased and ended just prior to implementation of sewage water treatment involving chemical precipitation with post-filtration. The third period (*c*) extended from the new waste treatment practices to just prior to implementation of new agricultural practices. The fourth period (*d*) comprised the period from when the newly implemented agricultural changes (which started a few years before 'Lex Ringsjön', a law in which Lake Ringsjön was considered to be a special pollution sensitive area, was introduced in 1985) were made until the fish reductions, and the fifth period (*e*) covered the time after the fish reductions, excluding the year following the fish reduction for each of the basins (Hansson et al., 1999). Only four periods were analysed for Sätofta Basin because it is upstream of the sewage water treatments plants, and thus was not directly affected by this measure (Table 1).

Results

Phosphorus loading

The phosphorus loadings between 1976 and 1980 were very high, ranging from 20 to 34 tons per year. I have no measurements from 1981 to 1983, but from 1984 and onwards the phosphorus loadings decreased and varied between 5 and 13 tons per year (Figure 1). In 1989 the phosphorus loading was as low as 4.7 tons. During the same year there was a pronounced increase in Secchi depth from a summer mean around 50–75 cm during the period 1976–1988, to over 100 cm in 1989 and afterwards (Figure 1). After 1989, the Secchi depth remained high even in years with rather high phosphorus loading (Figure 1).

Nutrients

In the late 1960s, the total-P concentration and PO_4-P concentration (summer means) were below 76 and 31 $\mu g\ l^{-1}$, respectively. The concentrations were lowest in Western Basin (Figure 2). In 1973 and after, the concentrations of total phosphorus increased dramatically in all three basins, and the concentration of PO_4-P increased in Sätofta Basin and Eastern Basin (Figure 2). From the mid-1980s, the amount of total-P decreased and a trend of decreasing total-P concentration was observed in all basins. Also PO_4-P decreased from the mid-1980s to levels similar to the period of 1966–1967. Although there is an increase in the P-loading in the 1990s, there is no increase in the concentration of phosphorus (total-P and PO_4-P) at this time, except in 1994–1996 in Eastern Basin (Figures 1 & 2). The higher levels of phosphorus in Eastern

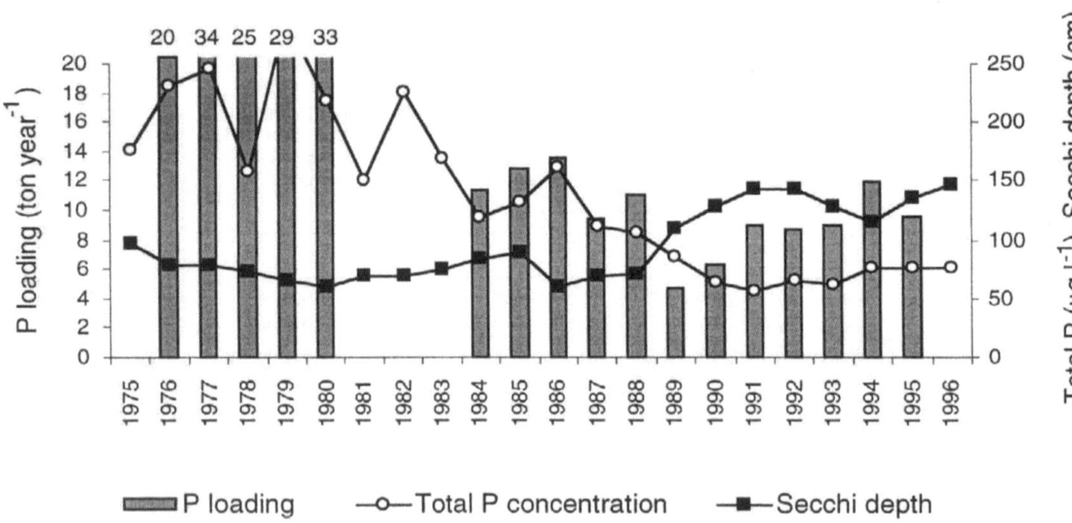

Figure 1. Annual P-loading (ton yr $^{-1}$, y_1-axis), total P-concentration (μg l^{-1}, y_2-axis), and Secchi depth (cm, y_2-axis) plotted against the year 1975–1996. The values of P-loading are written over the first five bars as they exceed the scale on the y-axis. Values for P-loading depict total loading on the whole lake, whereas total P-concentration and Secchi depth represent means of summer means for the three basins.

Basin are, however, probably due to strong early stratifications of the lake and oxygen depletion.

Three nitrogen fractions have been analysed, NO_2+NO_3-N, Kjeldahl-N (organic-N and ammonium-N) and total-N. In the 1960s, the concentrations of all nitrogen fractions in all basins were below 600 μg l^{-1}, and as observed for phosphorus, the lowest concentrations were found in Western Basin (Figures 2 & 3). Between 1973 and 1987, Kjeldahl nitrogen in all basins increased to levels between 1300 and 3000 μg l^{-1} (summer means). After 1987, the Kjeldahl-N stayed somewhat below 1500 μg l^{-1}. Total-N showed a pattern similar to Kjeldahl-N; between 1973 and 1987 (1985 for Eastern Basin) total nitrogen exhibited high variations between years with levels between 1300 and 4500 μg l^{-1} (summer means). After the mid-1980s, the total nitrogen decreased somewhat and has remained on levels between 1300 and 1900 μg l^{-1}. The concentration of NO_2+NO_3-N showed much less variation than Kjeldahl and total-N and through the whole period 1973–1994 it stayed below 660, 880 and 452 μg l^{-1} in Sätofta, Eastern and Western Basin, respectively (Figure 3).

Secchi depth, chlorophyll a *concentration and pH*

Secchi depth (summer means) has varied from 0.5 m to almost 3 m since 1966. In all three basins, the Secchi depth was around 2 m during the 1960s. In the early 1970s, Secchi depth decreased to levels around 0.5–1.0 m, but since 1989, 1988 and 1991 it has increased in Sätofta Basin and Eastern Basin and Western Basin, respectively. Since then Secchi depth has remained on levels above 1.0, 1.5 and 1.0 m (summer means) in Sätofta Basin, Eastern Basin and Western Basin, respectively (Figures 1 & 4).

Chlorophyll *a* was measured from 1975, and during the period 1975–1984 there was considerable variation, ranging from 35 to 175 μg l^{-1}. From 1985 to 1994, the between-year-variation was less pronounced, and instead, chlorophyll *a* concentrations decreased (Ln-transformed values, Regression, $F_{1,8}$=21.0, P=0.002 in Sätofta Basin, $F_{1,8}$=20.1, P=0.002 in Eastern Basin and $F_{1,8}$=36.2, P=0.0001 in Western Basin) (Figure 4). The summer means decreased from 51, 71 and 60 μg l^{-1} in 1984, to 30, 27 and 35 μg l^{-1} in 1994 in Sätofta, Eastern and Western Basins, respectively.

Prior to 1988, pH varied by one order of magnitude between years in Sätofta Basin, Eastern Basin and Western Basin. After 1988, variation between years

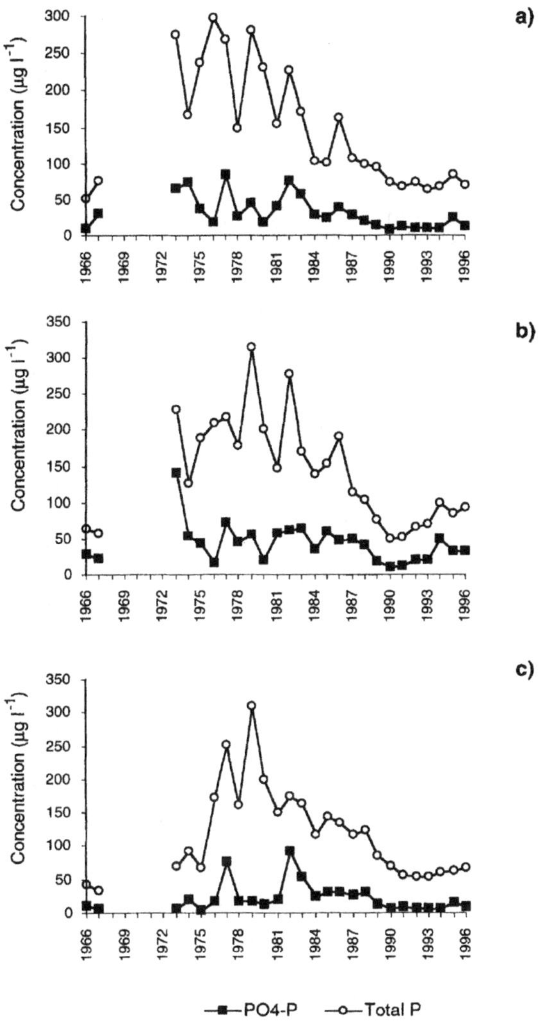

Figure 2. Summer means for PO_4-P ($\mu g\ l^{-1}$) and total-P ($\mu g\ l^{-1}$) plotted against the years 1966–1996 for: (a) Sätofta Basin; (b) Eastern Basin; and (c) Western Basin. Note that the *y*-axes have different scales.

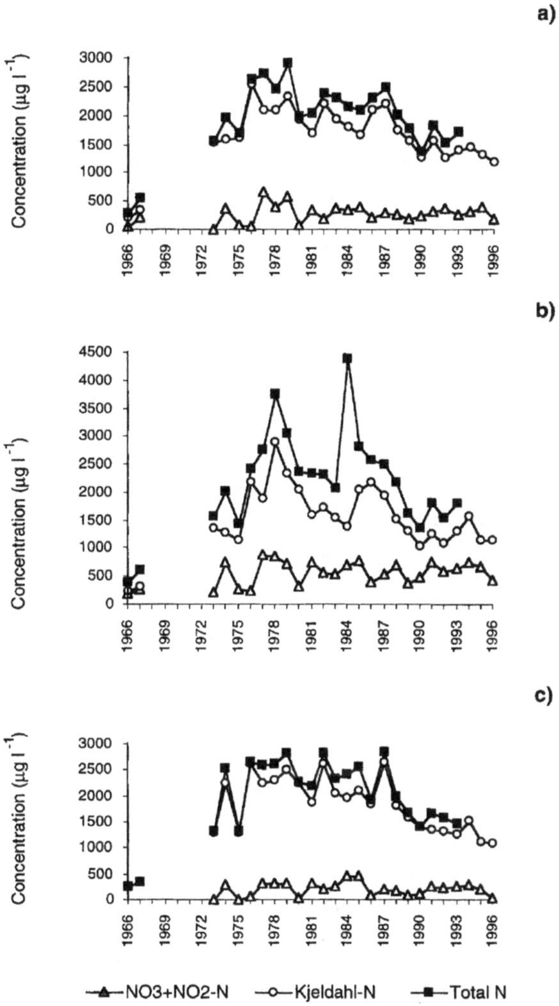

Figure 3. Summer means for NO_3+NO_2-N ($\mu g\ l^{-1}$), Kjeldahl-N ($\mu g\ l^{-1}$), and total nitrogen ($\mu g\ l^{-1}$) plotted against the years 1966–1996 for: (a) Sätofta Basin; (b) Eastern Basin; and (c) Western Basin. Note that the *y*-axes have different scales.

was smaller and pH averaged around 8.7 in all three basins (Figure 4).

Discriminant analysis (DA)

It is difficult to follow all of the changes in the lake as there are three different basins, each of which reacted differently to the changes in nutrient load and fish reductions. Consequently, I ran discriminant analyses (DA) to look at the whole data set simultaneously.

Separate DAs for each of the three basins showed statistically significant discrimination, based on water chemistry data (six variables) for the different time periods (Wilks' lambda, $F_{18,45}$=15.993, P=0.0001 for Sätofta Basin, Wilks' lambda, $F_{24,53}$=10.060, P=0.0001 for Eastern Basin, Wilks' lambda, $F_{24,53}$= 6.880, P=0.0001 for Western Basin). The DA assigned 96% of the 25 years in the analysis to the correct period for Sätofta Basin and Eastern Basin, and 88% of the years to the correct period for Western Basin. The pre-eutrophic years were greatly separated from all other periods, and appeared almost as outliers in the DA (Figure 5). For all three basins, the first two discriminant axes accounted for most of the discrimination. Kjeldahl-N had high loadings on the first discriminant axis for all three basins (>0.67), and Secchi depth had high loadings on the first discriminant axis for Sätofta Basin and Eastern Basin (>0.51). Total phosphorus had high loadings on the second axis for all three basins (>0.63), and Secchi depth had high loadings on the second axis for Western Basin (>0.6).

The discriminant analyses showed that the time periods of the different measures corresponded to observable changes in the lake water chemistry, primarily Secchi depth, total-P and Kjeldahl-N. During period *a* (1966–67), the lake was mesotrophic (Table 1, see also Hansson et al., 1999), which means low concentrations of both phosphorus and nitrogen, and a high Secchi depth. No measurements on chlorophyll *a* are available for this period (but see Cronberg, 1999). The first period was followed by a eutrophic period (Table 1), characterized by high phosphorus loading on the lake and consequently high phosphorus and nitrogen concentration in the lake. The high nutrient concentration also gave high chlorophyll *a* concentrations and low Secchi depths in all basins. Due to the very high nutrient concentrations during the early 1970s, efforts were undertaken to decrease nutrient loading on Lake Ringsjön. The two following time periods in the discriminant analysis corresponded to the improved sewage treatment and agricultural measures (Table 1, see also Hansson et al., 1999). The third period, after the new sewage treatment (Table 1), was nevertheless characterized by high nutrient loading, high nutrient concentrations, high or increasing chlorophyll *a* concentrations and low Secchi depths. During the fourth time period, after the new agricultural practices (period *d*; Table 1), phosphorus loading was much lower than during the 1970s, and consequently the total-P in the lake started to decrease (Enell & Fejes, 1998). There were no obvious changes in PO_4-P or the nitrogen concentration. The period after fish reduction was characterized by lower amplitudes and levels of the nutrient concentrations, lower chlorophyll *a* concentrations had higher Secchi depths. To summarize, Secchi depth was considerably greater in the first and fifth period than in the other periods, total phosphorus was greatest during periods 2–4, and Kjeldahl-nitrogen was greatest during periods 2–5.

Discussion

I present data obtained from numerous investigations of Lake Ringsjön that span a period of some 30 years (Björk & Lettervall, 1968; Ryding, 1983; Enell, 1985, 1986; Enell & Eriksson, 1987, 1988; Enell & Henriksson, 1989, 1990, 1991, 1992; Hamrin et al., 1991; Enell et al., 1993, 1994; Hamrin et al., 1993; Bergman et al., 1994; Enell & Fejes, 1996; Bergman, 1997). During these 30 years, the eutrophication process was reversed after implementing a number of measures, and the lake is less eutrophic today than it was in the 1970s and 1980s, even if it has not reached the former 'mesotrophic' state that it had in the early 1960s. Because several different measures have been undertaken, it is difficult to separate the effect of each of these measures on lake water chemistry. Nevertheless, I have found changes in water chemistry that seem to correspond to the different measures. Hence, improvements of the waste water treatment and agricultural practices were associated with the reduced phosphorus loading, and thus the lower nutrient level in the lake. The fish reductions helped maintain the low nutrient levels and also increased the Secchi depth, possibly by altering the algal community towards larger species (Cronberg, 1999).

The most important variables in the DA were Secchi depth, total-P and Kjeldahl-N concentrations, and Secchi depth was inversely related to the former variables. Total phosporus and Kjeldahl-N concentrations were more than three times as high during the second half of the 1970s compared to the mid 1960s. Several reasons for the eutrophication process may be noted:

(a) The number of people living in the area doubled during the 1950s, with increasing quantities of waste water containing nutrients as a consequence (Ljungström, 1970 in Ryding, 1983),
(b) The agricultural production was intensified, with increasing amounts of fertilizers as a consequence (Ryding, 1983; Hamrin et al., 1991),
(c) Recently, Enell & Fejes (1998) stated that the ecological effect of the water level regulation of Lake Ringsjön that started in 1963 may have affected the eutrophication process.

Investigations show that the regulation resulted in a highest water level occurring three months later than it did prior to 1963. Enell & Fejes (1998) also point out that the spring spawning fish probably were affected by this change.

One expectation following the new sewage treatment was that the nutrient concentrations in the lake

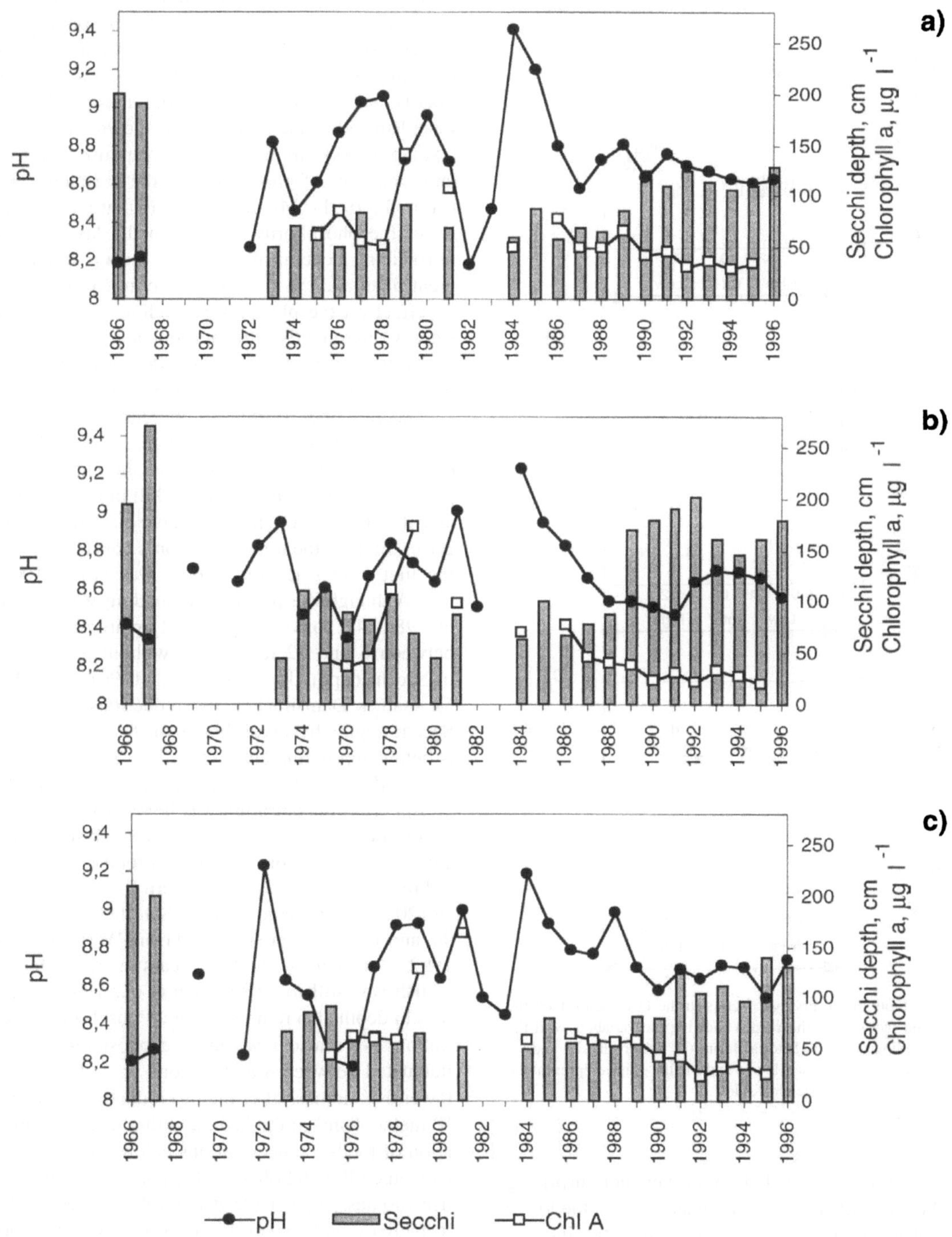

Figure 4. Summer means for pH (y_1-axis), Secchi depth (cm, y_2-axis), and chlorophyll *a* ($\mu g\ l^{-1}$) plotted against the years 1966–1996 for: (a) Sätofta Basin; (b) Eastern Basin; and (c) Western Basin. Note that the y_2-axes have different scales.

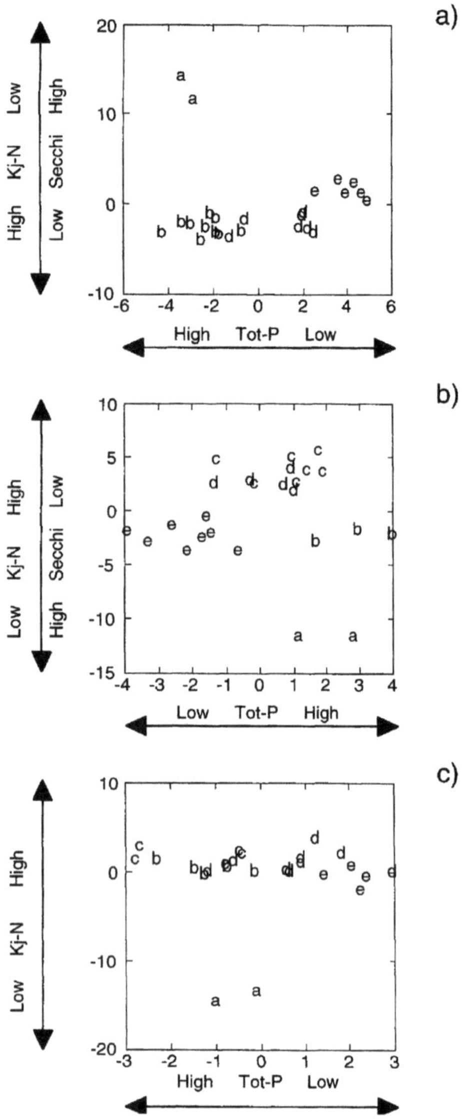

Figure 5. Plots of the first two factors in the DA (factor 1 on the *y*-axis, and factor 2 on the *x*-axis) showing the variables with the highest loadings, for: (a) Sätofta Basin; (b) Eastern Basin; and (c) Western Basin. The letters a, b, c, d and e in the graphs represent the different time periods listed in Table 1.

would again decrease. It is, however, not surprising that the lake did not respond immediately to the lower nutrient levels in the waste water, since the internal phosphorus loading was probably still high (Ryding, 1983; Enell & Fejes, 1998). Other eutrophic lakes in agricultural areas have had similar responses to decreases in nutrient loading (Boström, 1984; Björk, 1988; Sas, 1990). In lakes with low phosphorus loading and short residence times, up to ten years may be required before a lake responds to decreased phosphorus loading (Lean et al., 1990; Lathrop, 1990). For Lake Ringsjön, calculations have shown that it would take up to 50 years to observe a response, due to the large amounts of phosphorus in the sediment (Hamrin et al., 1991). My conclusion is that the efforts to lower the phosphorus loading on the lake, both by introducing a chemical step at the sewage water treatment plant and the fertilization control program, had an effect on the phosphorus loading on Lake Ringsjön. Consequently, the nutrient and chlorophyll *a* concentrations in the lake water decreased, which are necessary conditions to increase Secchi depth.

Secchi depth started to increase in 1989, co-occurring with three other changes. First, 1989 had the lowest phosphorus loading of the entire 30-year period. Second, there was a fish-kill in Eastern Basin, and the fish reduction programme started in Sätofta Basin. Finally, the total phosphorus concentration, for the first time since the 1960s, was below 100 μg l^{-1}. Although the phosphorus loading was very low in 1989 (4.7 tons), it increased again after 1989 to between 6 and 12 tons yr^{-1}, without any decrease in Secchi depth. Jeppesen et al. (1990) showed that for shallow Danish lakes, fish reductions can increase Secchi depths by more than 1 m if total phosphorus is below approximately 100 μg l^{-1}. This agrees with my results, which showed that after 1989, the total phosphorus concentration was below 100 μg l^{-1} and Secchi depth always stayed at levels above 1 m, regardless of phosphorus loading. A total of ca 500 tons of bream and roach died in Eastern Basin in 1988, and ca 120 tons were removed from Sätofta Basin in 1989–90, and 120 tons were removed from Western Basin in 1992. Between two and six years later, the biomass of fish was still low (Bergman et al., 1999), and the Secchi depth also remained high. Phosphorus concentrations were also low, and the phosphorus loading fluctuated between 8 and 13 tons yr^{-1}. I believe the lower fish biomass after the fish reduction affected the 'P metabolism' in the lake, and that it was a combination of total-P concentrations below 100 μg l^{-1} and a changed P metabolism in the lake, due to the lower fish biomass, that made it possible for Secchi depth to increase and remain high, even when phosphorus loading increased.

In summary, it appears that the addition of a chemical step in the sewage water treatment plant and the control program for farmers' usage of fertilizers had an effect on the nutrient levels in the lake. The fish reduction, in combination with the lower nutrient levels, seem to have had an effect on the Secchi depth. After the fish reduction, the lake seemed to be more tolerant to fluctuations in phosphorus loadings. The whole lake appears to have become a more stable ecosystem.

Acknowledgements

I thank Jonas Fejes at IVL, Stockholm, for providing me with computerized data files. Valuable comments on an earlier draft were given by Larry Greenberg and an anonymous reviewer. And thank you Daniel, my youngest child, for sleeping well while part of this manuscript was written.

References

Andersson, A., 1948. Näringstillgång och planktonutveckling i några skånska sjöar. Vattenhygien 1: 9–21 (in Swedish).

Almestrand, A. & A. Lundh, 1951. Studies on the vegetation and hydrochemistry of Scanian lakes. I–II. Botaniska Notiser Suppl. 2–3. Lund.

Bergman, E., E. Bergstrand, G. Cronberg, M. Eriksson & P. Romare, 1994. Sjörestaurering genom cyprinidreduktion. Ringsjöns status under 1993. Rapport. Limnologiska avdelningen, Lund (in Swedish).

Bergman, E., 1997. Ringsjön i Skåne, restaurering genom cyprinidreduktion – effekten av fiskreduktionen. Limnologiska avdelningen, Lunds Universitet. Rapport (in Swedish).

Bergman, E., S. F. Hamrin & P. Romare, 1999. The effects of cyprinid reduction on the fish community. Hydrobiologia 404: 65–75.

Björk, S., 1988. Redevelopment of lake ecosystems – a case-study approach. Ambio 17: 90–98.

Björk, S. & U. Lettervall, 1968. Bolmen - Lagan - Ringsjön. Resultat från en tvåårig limnologisk undersökning i samband med projektering för Skånes framtida vattenförsörjning. Sydvatten och Limnologiska institutionen, Lund Universitet. Rapport (in Swedish).

Boström, B., 1984. Potential mobility of phosphorus in different types of lake sediment. Int. Rev. ges. Hydrobiol. 69: 457–474.

Cronberg, G., 1999. Qualitative and quantitative investigations of phytoplankton in Lake Ringsjön, Scania, Sweden. Hydrobiologia 404: 27–40.

Enell, M., 1985. Ringsjön i förändring. Ringsjöns struktur och funktion 1984 samt förändring under perioden 1975–1984. Mellanskånes Planeringskommitté (MSK). 93 pp (in Swedish).

Enell, M., 1986. Ringsjön. Ekosystem i förändring 1985. Tabell- och figursammanställning för 1985 samt kortfattad beskrivning av sjöns struktur och funktion samt förändring under perioden 1975–1985. Mellanskånes planeringskommitté (MSK). 73 pp (in Swedish).

Enell, M. & M. Eriksson, 1987. Ringsjön. Ekosystem i förändring 1986. Tabell- och figursammanställning för 1986 samt kortfattad beskrivning av sjöns struktur och funktion samt förändring under perioden 1975–1986. Mellanskånes planeringskommitté (MSK). 82 pp (in Swedish).

Enell, M. & M. Eriksson, 1988. Ringsjön. Ekosystem i förändring 1987. Beskrivning av sjöns struktur och funktion samt förändring under perioderna 1984–1987 och 1975–1986. Mellanskånes planeringskommitté (MSK). 131 pp (in Swedish).

Enell, M. & J. Fejes, 1996. Ringsjön - ekosystem i förändring. Analys av tänkbara orsaker till sjöns eutrofiering. Stencil (in Swedish).

Enell, M. & J. Fejes, 1998. Exemplet Ringsjön - vad har vi lärt oss? Effekter i vatten. Stencil (in Swedish).

Enell, M., J. Fejes & A. Henriksson, 1993. Ringsjön. Ekosystem i förändring 1992. Beskrivning av sjöns struktur och funktion 1992 samt förändringar under 1975–1992. Mellanskånes kommunalförbund. 111 pp (in Swedish).

Enell, M., J. Fejes & A. Henriksson, 1994. Ringsjön. Ekosystem i förändring 1991–1993. Beskrivning av sjöns struktur och funktion 1991–1993 samt förändringar under 1975–1993. Mellanskånes kommunalförbund. 153 pp (in Swedish).

Enell, M. & J. Henriksson, 1989. Ringsjön. Ekosystem i förändring 1988. Beskrivning av sjöns struktur och funktion 1988 samt förändringar under 1975–1988. Mellanskånes kommunalförbund. 139 pp (in Swedish).

Enell, M. & J. Henriksson, 1990. Ringsjön. Ekosystem i förändring. Beskrivning av sjöns struktur och funktion 1989 samt förändringar under 1975–1989. Mellanskånes kommunalförbund. 90 pp (in Swedish).

Enell, M. & J. Henriksson, 1991. Ringsjön. Ekosystem i förändring. Beskrivning av sjöns struktur och funktion 1990 samt förändringar under 1975–1990. Mellanskånes kommunalförbund. 133 pp (in Swedish).

Enell, M. & J. Henriksson Fejes, 1992. Ringsjön. Ekosystem i förändring. Beskrivning av sjöns struktur och funktion 1991 samt förändringar under 1975–1991. Mellanskånes kommunalförbund. 125 pp (in Swedish).

Hamrin, S. F., 1999. Planning and execution of the fish reduction in Lake Ringsjön. Hydrobiologia 404: 59–63.

Hamrin, S. F., E. Bergstrand, G. Cronberg & P. Romare, 1991. Sjörestaurering genom cyprinidreduktion. Ringsjöns utveckling sedan 1966 samt effekter av fiskreduktionen 1988–1990. Limnologiska avdelningen. Lund. Rapport (in Swedish).

Hamrin, S. F., E. Bergstrand, G. Cronberg, A. Persson, P. Romare & J. Strand, 1993. Sjörestaurering genom cyprinidreduktion. Ringsjöns utveckling under perioden 1987–1992. Stencil. Limnologiska avdelningen, Lund och Fiskeriverkets Sötvattenslaboratorium, Drottningeholm (in Swedish).

Hansson, L.-A., M. Enell & E. Bergman, 1999. Lake Ringsjön: its catchment area, its history and its importance. Hydrobiologia 404: 1–7.

Jeppesen, E., J. P. Jensen, P. Kristensen, M. Søndergaard, E. Mortensen, O. Sortkjær & K. Olrik, 1990 . Fish manipulation as a lake restoration tool in shallow, eutrophic temperate lakes 2: threshold levels, long-term stability and conclusions. Hydrobiologia 200/201: 219–227.

Lathrop, R. C., 1990. Response of Lake Mendota (Wisconsin, U.S.A.) to decreased phosphorus loadings and the downstream lakes. Verh. Int. Ver. Limnol. 24: 457–463.

Lean, D. R. S., M. A. Neilson, R. J. J. Stevens & A. Mazumder, 1990. Response of Lake Ontario to reduced phosphorus loading. Verh. Int. Ver. Limnol. 24: 420–425.

Ljungström, K., 1970. Ringsjöns eutrofiering till följd av ökande föroreningsbelastning – återblick och prognostiska överväganden. Vatten 26: 184–193 (in Swedish).

Lundh, A., 1951a. Studies on the vegetation and hydrochemistry of Scanian lakes. III. Distribution of macrophytes and some algal groups. Botaniska Notiser Suppl. 3: 1.

Lundh, A., 1951b. Some aspects of the higher aquatic vegetation in the Lake Ringsjön In Scania. Botaniska Notiser 1: 21–31.

Ryding, S.-O., 1978. Recovery of polluted lakes. Loading, water quality and responses to nutrient reduction. Acta Univ. Upsal., Abstracts of Uppsala dissertations, Faculty of Science. Nr 459.

Ryding, S.-O., 1983. Ringsjöområdet - ekosystem i förändring. Redovisning och utvärdering av Limnologiska undersökningar. Mellanskånes planeringskommittee och Limnologiska institutionen, Uppsala Universitet. Januari 1983 (in Swedish).

Sas, H., 1990. Lake restoration by reduction of nutrient loading: Expectations, experiences, extrapolations. Verh. Int. Ver. Limnol. 24: 247–251.

Thorntwaite, C. W., 1948. The determination of potential evaporation. John Hopkins Univ. Lab., Climat., Vol. III, No. 1.

Trybom, F., 1893. Ringsjön i Malmöhus län dess naturförhållande och fiske. Medd. Kungl. Lantbruksstyr. 4: 1–42 (in Swedish).

Hydrobiologia **404:** 19–26, 1999.
L.-A. Hansson & E. Bergman (eds), Nutrient Reduction and Biomanipulation as Tools to Improve Water Quality: The Lake Ringsjön Story.

Internal phosphorus loading in Lake Ringsjön

Wilhelm Granéli
Department of Limnology, Insitute of Ecology, Ecology Building, University of Lund, SE-223 62 Lund, Sweden
E-mail: wilhelm.graneli@limnol.lu.se; Tel; +[46] 46 2228431; Fax +[46] 46 2224536

Key words: sediment, phosphorus, eutrophication, biomanipulation, fish, Sweden

Abstract

Lake Ringsjön did not respond with decreased algal production following a substantial reduction in external phosphorus loading. This is typical of many shallow lakes which for decades have received excessive amounts of nutrients. The inertia is due to large amounts of phosphorus (P) stored in sediments and biota, causing internal phosphorus loading. Much of this phosphorus is thought to be released from the organic-rich profundal sediments. In Lake Ringsjön, only one third of the total bottom area is covered by such sediments, the rest being dominated by sand and silt. In the profundal sediments bulk P content was not exceptionally high (approximately 2 mg P·g DW^{-1}), while the pore water phosphate concentrations, especially in Sätofta Basin, were very high, indicating large potential for phosphorus release to the water. This is also indicated by the large proportion of Fe- and Al-bound P in the sediments of Sätofta Basin. Although there are no direct quantifications of phosphorus release from the sediments in Lake Ringsjön, measurements of phosphorus concentrations in the water mass as well as budget calculations for the three basins clearly show a high capacity for internal loading. Phosphorus concentrations generally increase during summer, when external additions are minimal. Until 1980, the annual external phosphorus addition to Lake Ringsjön greatly exceeded the output, showing that the lake was an efficient phosphorus trap. Since then, input and output have been balanced, but in recent years signs that the lake is once again retaining phosphorus on an annual basis are evident. There are marked differences between the three basins, with Western Basin generally retaining phosphorus, while the upstream Eastern Basin and Sätofta Basin during the 1980s often exported phosphorus. It is not possible to evaluate the effects of the fish biomanipulation on the internal loading of phosphorus from the sediment.

Introduction

Like with many other lakes, Lake Ringsjön has become hypertrophic due to decades of excessive increases in the input of nutrients, through domestic sewage and agricultural run-off. In many cases, hypertrophic lakes do not respond quickly (within a few years) to even a drastic reduction in the external nutrient loading (Cullen & Forsberg, 1988). Massive algal blooms, usually of cyanophytes, may continue to occur for many years or even decades. This is because lakes normally are efficient nutrient traps, where much of the external nutrient load is accumulated (Kamp-Nielsen, 1997). These nutrients buffer the ecosystem and cause a more or less pronounced inertia with respect to effects from changes in the external nutrient loading. Large amounts of phosphorus are found in the biota, such as in fish, plankton and macrophytes (Andersson et al., 1988; Granéli & Solander, 1988). This phosphorus may be recycled on rather short time scales (hours–months). However, most external phosphorus input that has not left the lake through outflow is found in the sediments, especially in the highly organic sediments of the profundal parts of the lake. The pool of nitrogen and phosphorus in the top 10 cm of profundal sediments may represent many years of external loading. Part of the phosphorus accumulated in the sediments can be mobilized to the overlying water in an algal-available form. The magnitude of this sediment–water exchange, and the biotic and abiotic factors controlling it, have been important issues in both basic and applied limnological research, especially during the sixties and seventies (Boström et al., 1982; Enell & Löfgren,

1988; Boström et al., 1988). Phosphorus is bound to the sediment in several different forms (Holtan et al., 1988). The proportions of these forms vary over a wide range between lakes. Consequently, sediment–water exchange of phosphorus may be regulated through different mechanisms in different lakes. It is generally recognized that oxygen, iron and pH are important chemical regulating factors, but biological mechanisms, e.g. bacterial uptake and excretion, as well as bioturbation by benthic macrofauna, including fish have also been emphasized (Boström et al., 1982; Kairesalo et al., 1995). In fact, biological and chemical factors cannot be separated. High pH may promote leaching of phosphorus from the sediment, but high pH in eutrophicated lakes is partly the result of high pelagic primary production, which in turn is ultimately caused by high (external) loading of nutrients, since nutrient availability, and particularly phosphorus, often limits primary production in lakes. High primary production and a high algal biomass may also be connected to heavy predation by planktivorous fish on larger zooplankton, whereby herbivory is reduced. Furthermore, algal blooms, when they crash, will lower redox potentials in the surficial sediment layer due to excessive sedimentation of degradable organic matter, which may promote leaching of sediment phosphorus. A shallow lake, like Lake Ringsjön, is vulnerable with respect to internal loading, because nutrients stored in sediments have a greater impact on water quality than in deeper lakes.

As early as several decades ago, it was realized that reduction in internal nutrient loading may be required to return a lake to a less eutrophic state. A reduction in external loading, to levels where phosphorus should theoretically be limiting algal production, may be counteracted by internal loading for many years (Björk et al., 1979). Starting in the late 1960s, sediment dredging, hypolimnetic aeration and chemical sediment phosphorus inactivation have been tested in many lakes with varying success (Cooke et al., 1993). Successful biomanipulation of a hypertrophic lake may be directly (fish bioturbation) or indirectly (reduction in pH and sedimentation) connected to inactivation of the sediment store of phosphorus. It may also be a pure food-chain effect according to the classical model, where a reduction in planktivorous fish causes herbivory on phytoplankton to increase (Bergman et al., 1999).

In the presentation below, I use results from sediment investigations made in Lake Ringsjön in 1981–82 (Enell, 1983). I discuss the role of the sediments in internal phosphorus loading, using budget calculations for the whole lake as well as for the three sub-basins separately. I have focused on phosphorus, since this element is generally thought to be the key factor in eutrophication of lakes. Sediments are also important with respect to nitrogen cycling however (see Svensson et al., 1999), which may influence the N:P ratio and thus the balance between different groups of phytoplankton (Schöllhorn & Granéli, 1997).

Study site

Lake Ringsjön is situated in southernmost Sweden (55° 52′ N; 13° 32′ O) and consists of three basins: Sätofta Basin (4.2 km^2, mean depth 3.0 m, maximum depth 17.0 m), Eastern Basin (20.5 km^2, mean depth 6.1 m, maximum depth 16.4 m) and Western Basin (14.8 km^2, mean depth 3.1 m, maximum depth 5.4 m)(Hansson et al., 1999). As with many other lakes in urban and agricultural areas, Lake Ringsjön became eutrophicated during the 1960s and 1970s. In the beginning of the 1980s, a *nutrient reduction programme* was initiated which reduced the external input of nutrients to the lake from phosphorus loadings exceeding 30 tons per year to around 10 tons per year. About ten years later (1989–1990), a *cyprinid fish reduction programme* (biomanipulation) was carried out in Sätofta Basin removing about 100 tons of fish (about 60% of the standing stock). During the winter of 1988, there was a fish-kill in Eastern Basin, eliminating about 500 tons of fish (about 80%) (Hamrin, 1999). The biomanipulation in Western Basin, which was performed in 1992, removed about 120 tons of fish (about 50%).

Sediment investigations in Lake Ringsjön

Lake Ringsjön sediments were studied in 1981–1982, when the lake was near its most eutrophic state (Secchi depth had its lowest values during this century at the end of the seventies (Verner, 1994)). Investigations included horizontal distribution of organic sediments, their thickness and total content of nutrients, heavy metals, pesticides and organic matter (Enell, 1983). The vertical distribution of nutrients, both in bulk sediment and in sediment pore water, was also studied. Phosphorus fractionation studies were made on surficial sediment. Several sediment cores were dated, making it possible to roughly calculate sediment growth rates and rates of accumulation of

Table 1. Sediment characteristics for the three basins of Lake Ringsjön. Based on data from Enell (1983)

Basin	Org. Sed.[1]	Sed. growth[2]	Sediment P[3]	Interstitial P[4]	Sed. P exchange[5]	
	%	mm yr^{-1}	mg P g^{-1} DW	mg P l^{-1}	max mg P m^{-2} d^{-1}	mean
Sätofta	27	—	1.89	2.2	11.7	3.5
Eastern	43	2.3	2.32	0.8	33.7	10.9
Western	32	3.3	1.74	0.4	7.1	4.0

[1] Defined by echo-sounding.
[2] Based on wet sediment.
[3] In bulk sediment.
[4] Dissolved inorganic phosphate.
[5] Net release.

nutrients. Separate investigations were carried out for the three sub-basins. Unfortunately, there are no results available either for the period before the peak in eutrophication or after the fish reduction experiment was completed. In addition, no experimental studies of sediment–water exchange of nutrients or of sediment oxygen uptake were performed. There are however, phosphorus and nitrogen budgets for the three sub-basins as well as for the whole lake, covering a long time period (1984–1996), including the post-biomanipulation period, but not the 'pristine' stage of the lake (Ryding, 1983a,b; Enell, 1985, 1986; Enell & Eriksson, 1987, 1988; Enell & Henriksson, 1989, 1990, 1991; Enell & Henriksson-Fejes, 1992; Enell et al., 1993, 1994; Hamrin et al., 1993; Enell & Fejes, 1995; Enell et al., 1996, 1997; Bergman, 1997). These budget calculations also cover seasonal variations. Thus, it is possible to determine if the whole lake complex or its various sub-basins function as nutrient traps or as net sources of phosphorus over the year. It is also possible to draw tentative conclusions concerning seasonal variation in internal loading of phosphorus. Budget calculations however, have inherent uncertainties due to the short hydrologic residence time of the basins and the lack of ground-water data. Nutrient budgets also do not reveal the source of internal loading, whether it is profundal sediments, littoral sediments or excretion/mineralization from fish or other biota, which are generally not included in standard water total phosphorus analyses.

Distribution and composition of profundal sediments

All three basins of Lake Ringsjön are functionally shallow, with mean depths similar to or less than the depth of the epilimnion during summer. Since the basins are not normally vertically stratified, it can be assumed that wind redistribution of sedimenting material is effective, especially since the surroundings are comparatively flat and wind-exposed. Thus, profundal sediments, i.e. highly organic and fine grained deposits, only cover part of the basins. The upper depth limits for organogenic sediments are similar to the mean depths of the basins. The fractions of the total bottom areas covered by organogenic sediments (based on echo-sounding) are 27%, 43% and 32% for Sätofta Basin, Eastern Basin and Western Basin, respectively (Table 1). Large areas of the lake bottoms are covered by sand or silt (63% for the whole of Lake Ringsjön), where no permanent deposition of organogenic sediments occurs. Even in the deeper parts of the lake there are areas devoid of such sediments. Maximum sediment depth is 5 m, but in most cases the organogenic sediment layer is much thinner.

In the surficial sediment (0–2 cm) all three basins showed similar phosphorus concentrations (Table 1, Figure 1), with somewhat higher values in Eastern Basin. With respect to phosphate in the interstitial water (average for 6 levels 0–20 cm and 6 sampling occasions autumn 1981 – spring 1982) Sätofta Basin showed higher values than the other two basins (Table 1). The total phosphorus content of the sediment (mg P·g DW^{-1}) is not exceptionally high, but the phosphate levels in the interstitial water indicate eutrophic to hypertrophic conditions and a large potential for internal phosphorus loading from these sediments (Enell & Löfgren, 1988).

There were marked differences between the three basins with respect to phosphorus fractions in surficial sediments (Ryding, 1983a,b). Sätofta Basin had a much larger proportion of Al- and Fe-bound phosphorus (fractions were operationally defined according to Hieltjes & Lijklema (1980)) compared to the other two basins (Figure 1). In Eastern Basin and

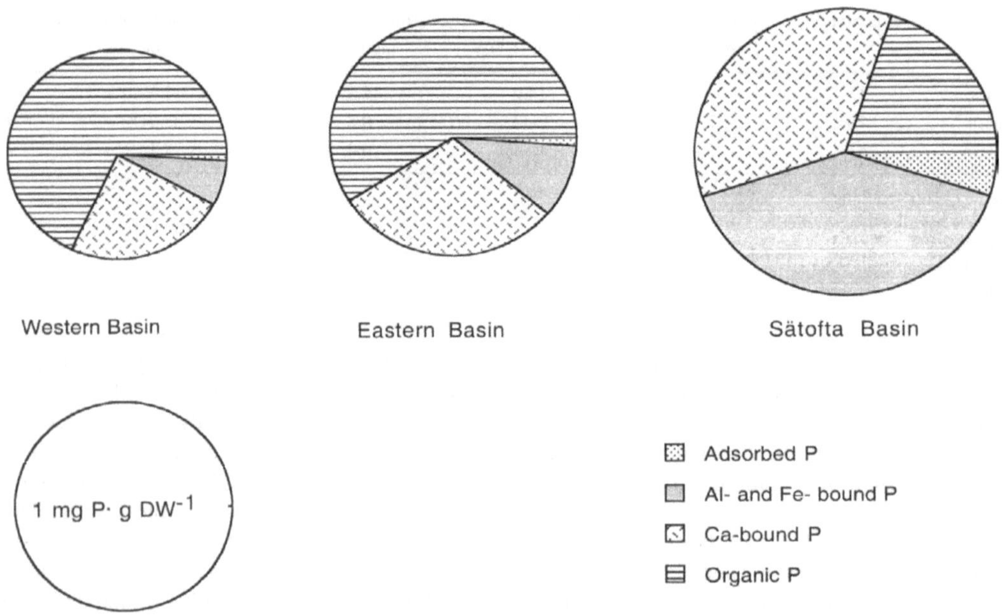

Figure 1. Phosphorus fractions in surface sediment (0–2 cm) from the three basins of Lake Ringsjön. Redrawn from Ryding (1983). Fractionation is according to Hieltjes & Lijklema (1980).

Western Basin organic bound phosphorus dominated, while the proportion of Ca-bound phosphorus was similar for all three basins. Unfortunately it is difficult to interpret these figures in terms of phosphorus release, since there may be no straightforward connection between the operationally defined sediment phosphorus fractions and mobility of the phosphorus under various environmental conditions (pH, redox potential, temperature) (Holtan et al., 1988). Fe-bound phosphorus is thought to be mobilized under reducing conditions, which may be more common in Sätofta Basin. However, sediment–water exchange rates for phosphorus, based on budget calculations, were higher in Eastern Basin (Table 1), indicating that other mechanisms may be operating, e.g. release of sediment phosphorus due to high pH in the overlying water.

Sediment growth rate

The rate of sediment accumulation has been estimated through ^{210}Pb and ^{137}C -dating of the top 100 cm of sediment cores. Both methods gave similar results with respect to sediment accumulation during the last 100 years; for Eastern Basin 2.3 mm yr^{-1} (wet sediment) and for Western Basin 3.3 mm yr^{-1} (Table 1). For Sätofta Basin dating was unsuccessful, indicating substantial sediment mixing. The above reported rates of sediment accumulation are representative for the deepest parts of the basins. At the end of the 19th century the water level of Lake Ringsjön was lowered by 1.5 m, causing a substantial increase in sedimentation due to erosion and subsequent redistribution of sediments. Before the water table was lowered, sedimentation was estimated to have been 0.5–1.5 mm yr^{-1}, a range indicating mesotrophic conditions (Björk et al.,1979; Enell, 1983). The rate of net sediment accumulation estimated for Lake Ringsjön for the decades before 1982, although being higher than during the 19th century, does not seem to be consistent with the hypertrophic state of the lake. The explanation of this may be the low mean depth and high wind exposure, causing frequent sediment resuspension and possibly export of particulate material downstream.

In lakes receiving high loads of domestic sewage, it is often possible to distinguish a surficial sediment layer with increased concentrations of N and P. This layer is sometimes colored black due to the forma-

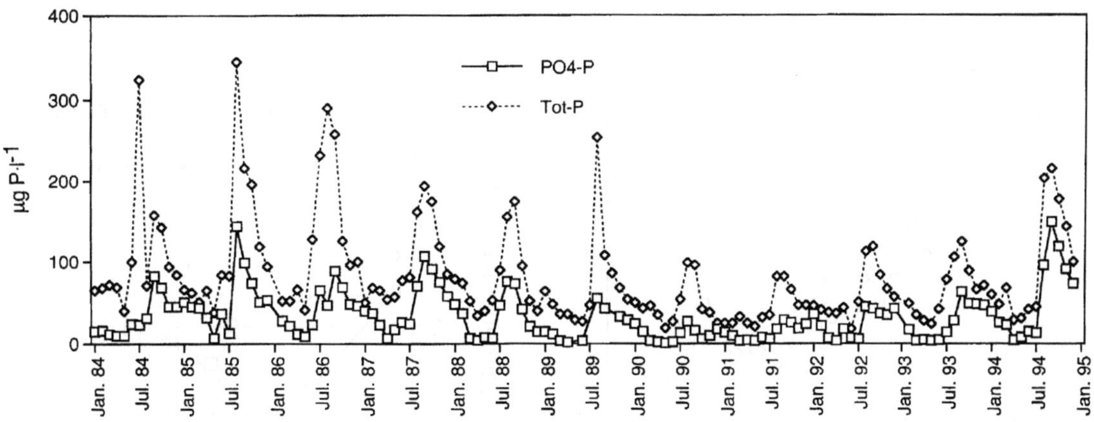

Figure 2. Water column mean phosphate and total-P concentrations in Eastern Basin 1984–1994.

tion of iron sulfide formed under reducing conditions. The rate of sediment accumulation is markedly higher than it was before the lake became a recipient. Gas formation is often intense in these 'recipient' sediment layers. A recipient sediment layer was only noticed in Sätofta Basin (Enell, 1983). However, since no dating of the sediment core was possible, nothing can be said about the time period of the formation of the layer. In Eastern Basin and Western Basin, a somewhat increased content of phosphorus can be noted in the upper 35–40 cm of sediment cores from the deeper part of the basins (Enell, 1983). If the estimated rates of 20th century sediment accumulation are correct, this would represent more than 100 years of sedimentation. Thus, the accumulation of a sediment layer with elevated phosphorus content seems to have started well before the period when the lake became hypertrophic (1960s).

Internal phosphorus loading

As has been noted above there are no studies in Lake Ringsjön revealing capacity for sediment accumulation or release of phosphorus. Neither can it be said with certainty under what environmental conditions (pH, redox, oxygen, bioturbation) a release would be maximal. There are however, strong indications of a substantial internal loading of the lake water with phosphorus. The organogenic sediments are one likely source of this phosphorus. A clear sign of internal fertilization is an increase in total phosphorus and dissolved inorganic phosphate in the water mass during the course of the summer. Typically, phosphorus (both total phosphorus and phosphate) shows a rapid increase from May to August, followed by an equally rapid decrease in autumn-winter (Figure 2). The changes are similar for surface, intermediate and deep water, illustrating that Lake Ringsjön is functionally shallow and comparatively well mixed during much of the year. In recent years (1990 and onwards) the summer increase in phosphorus has been less dramatic, indicating reduced internal loading. The pattern of increasing phosphorus concentrations during summer is found in several hypertrophic, shallow lakes in southern Sweden (e.g. Hamrin et al., 1998) and may be characteristic of hypertrophic lakes in general. A summer increase in phosphorus is contrary to what would be expected, at least for dissolved inorganic phosphate, which in less phosphorus-rich lakes is almost totally consumed during the vegetation period. The increase in phosphorus during summer coincides with low external loading. Using budget calculations (based on measurements of input, output and changes in the phosphorus pool in lake water), it becomes evident that there must be a substantial internal phosphorus source in Lake Ringsjön. The increase in phosphorus during summer coincides with a marked increase in pH, often reaching 9.5 – 10 in late summer.

Until 1980, the input of phosphorus to Lake Ringsjön greatly exceeded the export through outflowing water from the lake (Enell & Fejes, 1995) (Figure 3). Thereafter, input and output have been more balanced, and for some years there has even been a greater export than input of phosphorus to the lake system. In recent years, Lake Ringsjön has once again begun to function as a phosphorus trap. Lakes normally accumulate phosphorus. Thus, external loading

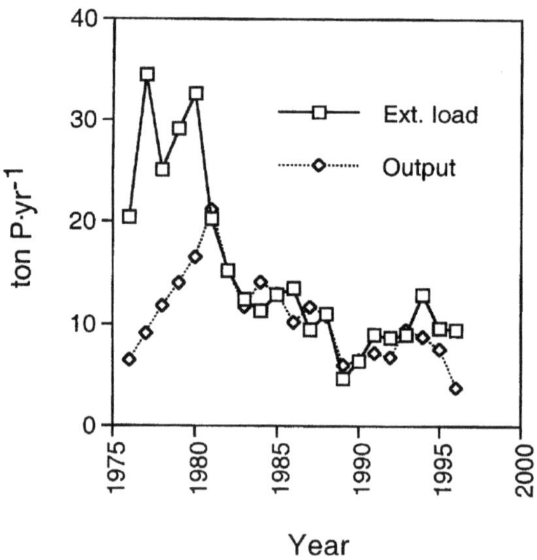

Figure 3. Annual external loading to and output of phosphorus from Lake Ringsjön.

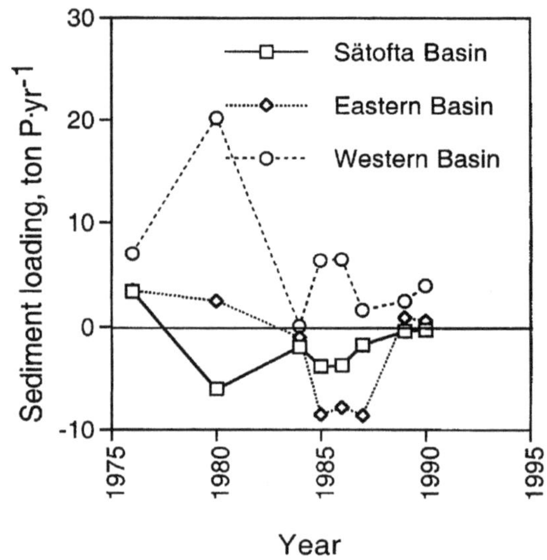

Figure 4. Net annual phosphorus accumulation (positive values) or release (negative values) in the three basins of Lake Ringsjön.

in most lakes greatly exceeds the export of phosphorus and a substantial fraction of the input is accumulated in the sediments. A prolonged (several years) export of phosphorus greater than the import (taking into account changes in the pool of phosphorus in the total water mass) indicates transient conditions caused by a long-term reduction in phosphorus input. Such a disequilibrium cannot of course continue indefinitely, but given the large supply of phosphorus that can accumulate in the sediments during the recipient period, the time scale may be in the order of decades. Even if seasonal internal loading is large, losses through outflow may still be small since internal phosphorus loading is at its maximum in summer, when the inflow and outflow of water is minimal. In the autumn, before the lake is flushed due to high late autumn and winter run-off, phosphorus concentrations are markedly lowered. Hypertrophic lakes, especially if external nutrient loading has been reduced, often show low or negative phosphorus retention, in contrast to oligotrophic and mesotrophic lakes (Ahlgren et al., 1979; Björk et al., 1979).

Judging from the annual phosphorus budgets, the internal net phosphorus loading of Lake Ringsjön has been small since measurements began in the mid-seventies (Figure 3). If the basins are treated separately however, it is seen that Western Basin, which receives its water from the other two basins, has since 1975 generally functioned as a phosphorus trap (Figure 4). Eastern Basin exported large amounts of phosphorus in 1985 and 1986, but has then moved towards a balance between loading and export. Also Sätofta Basin has changed from a phosphorus trap to a source of phosphorus, but since the late eighties, annual internal net loading from this basin is small.

Internal loading of phosphorus at the end of the seventies, calculated per surface area and time, was highest in Eastern Basin and appreciably lower in the other two basins (Table 1, see also Ryding, 1983a,b). The maximal rates are typical of hypertrophic lakes, although by no means extreme (cf. van der Molen & Boers, 1994). In laboratory experiments, similar rates have been reported for hypertrophic lake sediments, especially under anoxic conditions (Ahlgren et al., 1979; Granéli, 1979; Kamp-Nielsen, 1997).

Even though there may not be any large net mobilization of phosphorus from Lake Ringsjön on an annual basis, the fact that the output of phosphorus is still similar to the input indicates that the lake is not in equilibrium. If there was not such a large internal store of phosphorus in the lake, output would be markedly lower than input. There are additional indications of the existence of internal phosphorus loading in Lake Ringsjön. The increase during summer of the phosphorus content in the water has been mentioned above. When the lake was at its peak of hypertrophication, the external input of phosphorus should, theoretically, have resulted in a mean water column total phosphorus concentration of approximately 100 $\mu g\ l^{-1}$ in summer. Due to internal phosphorus loading, the concentration was 200–300 $\mu g\ l^{-1}$ in the three basins (Ryding, 1983a,b).

Concluding remarks

Lake Ringsjön has gone through a long period of eutrophication caused by greatly increased external nutrient loading. The external loading of phosphorus has been substantially reduced over the past two decades. As a result of this reduction in external loading the lake system has entered a transient phase of net phosphorus export, presumably of phosphorus stored in the sediments during the recipient period. The lake has thereafter returned to functioning as a phosphorus trap, though not yet a very efficient one. To what extent fish reduction has influenced internal (sediment) phosphorus loading is difficult to evaluate. As pointed out by Phillips et al. (1994), fish removal may cause an increase in abundances of benthic invertebrates, such as in bioturbating chironomid larvae, thus increasing the transport of phosphorus from sediment to water (Andersson et al., 1988, Svensson et al., 1999). Conversely, fish removal may, through cascading trophic interactions or through changes in excretion of phosphorus, cause changes in pH, oxygen and submersed vegetation, leading to decreased leakage of sediment phosphorus to the overlying water. Hence, the net effect of fish removal on internal P-load is difficult to predict.

References

Ahlgren, I. et al., 1979. Lake metabolism – Studies and results at the institute of limnology in Uppsala. Arch. Hydrobiol. Beih. Ergebn. Limnol. 13: 10–30.

Andersson, G., W. Granéli & J. Stenson, 1988. The influence of animals on phosphorus cycling in lake ecosystems. Hydrobiologia 170: 267–284.

Bergman, E., 1997. Ringsjön i Skåne – restaurering genom cyprinidreduktion – effekter av fiskreduktionen (Lake Ringsjön in Scania – rehabilitation through cyprinid reduction – effects from the fish reduction). Dept. of Limnology, Lund University, 77 pp. (In Swedish).

Bergman, E., L.-A. Hansson & G. Andersson, 1999. Biomanipulation in a theoretical and historical perspective. Hydrobiologia 404: 53–58.

Björk, S. et al. 1979. Lake management – Studies and results at the institute of limnology in Lund. Arch. Hydrobiol. Beih. Ergebn. Limnol. 13: 31–55.

Boström, B., J. M. Andersen, S. Fleischer & M. Jansson, 1988. Exchange of phosphorus across the sediment–water interface. Hydrobiologia 170: 229–244.

Boström, B., M. Jansson & C. Forsberg, 1982. Phosphorus release from lake sediments. Arch. Hydrobiol. Beih. Ergebn. Limnol. 18: 5–59.

Cooke, G. D., E. B. Welch, S. A. Peterson & P. R. Newroth, 1993. Restoration and management of lakes and reservoirs, second edition. Lewis Publishers, Boca Raton, 548 pp.

Cullen, P. & C. Forsberg, 1988. Experiences with reducing point sources of phosphorus to lakes. Hydrobiologia 170: 321–336.

van der Molen, D. T. & P. C. M. Boers, 1994. Influence of internal loading on phosphorus concentration in shallow lakes before and after reduction of the external loading. Hydrobiologia 275/276: 379–389.

Enell, M., 1983. Sedimentens betydelse i Ringsjöns eutrofiering (The importance of the sediments for the eutrophication of Lake Ringsjön). Institute of Limnology, University of Lund, 110 pp. (In Swedish).

Enell, M., 1985. Ringsjön – ekosystem i förändring 1984: Ringsjöns struktur och funktion 1984 samt förändringar under perioden 1975–1984 (Lake Ringsjön – Ecosystem under change 1984: The structure and function of Lake Ringsjön 1984 and changes during the period 1975–1984). Mellanskånes Planeringskommitté and Dept. of Limnology, Lund University, 94 + 15 pp. (In Swedish).

Enell, M., 1986. Ringsjön – ekosystem i förändring 1985: Tabell- och figursammanställning för 1985 samt kortfattad beskrivning av sjöns struktur och funktion samt förändringar under perioden 1975–1985 (Lake Ringsjön – Ecosystem under change 1985: Tables and Figures for 1985 and a short description of the structure and function of Lake Ringsjön and changes during the period 1975–1985). Mellanskånes Planeringskommitté and Institute of Water and Air research (IVL), 73 pp. (In Swedish).

Enell, M. & M. Eriksson, 1987. Ringsjön – ekosystem i förändring 1986: Tabell- och figursammanställning för 1986 samt kortfattad beskrivning av sjöns struktur och funktion samt förändringar under perioden 1975–1986 (Lake Ringsjön – Ecosystem under change 1986: Tables and figures for 1986 and a short description of the structure and function of Lake Ringsjön and changes during the period 1975–1986). Mellanskånes Planeringskommitté and Institute of Water and Air research (IVL), 82 pp. (In Swedish).

Enell, M. & M. Eriksson, 1988. Ringsjön – ekosystem i förändring 1987 – Beskrivning av sjöns struktur och funktion 1987 samt förändringar under 1975–1987 (Lake Ringsjön – Ecosystem under change 1987 – Description of the structure and function of Lake Ringsjön 1987 and changes during 1975–1987). Mellanskånes Planeringskommitté and Institute of Water and Air research (IVL), 104 pp. (In Swedish).

Enell, M. & J. Henriksson, 1989. Ringsjön – ekosystem i förändring 1988 – Beskrivning av sjöns struktur och funktion 1988 samt förändringar under 1975–1988 (Lake Ringsjön – Ecosystem under change 1988 – Description of the structure and function of Lake Ringsjön 1988 and changes during 1975–1988). Mellanskånes Planeringskommitté and Institute of Water and Air research (IVL), 138 pp. (In Swedish).

Enell, M. & J. Henriksson, 1990. Ringsjön – ekosystem i förändring 1989 – Beskrivning av sjöns struktur och funktion 1989 samt förändringar under 1975–1989 (Lake Ringsjön – Ecosystem under change 1989 – Description of the structure and function of Lake Ringsjön 1989 and changes during 1975–1989). Mellanskånes Kommunalförbund and Institute of Water and Air research (IVL), 138 pp. (In Swedish).

Enell, M. & J. Henriksson, 1991. Ringsjön – ekosystem i förändring 1990 – Beskrivning av sjöns struktur och funktion 1990 samt förändringar under 1975–1990 (Lake Ringsjön – Ecosystem under change 1990 – Description of the structure and function of Lake Ringsjön 1990 and changes during 1975–1990). Mellanskånes Kommunalförbund and Institute of Water and Air research (IVL), 133 pp. (In Swedish).

Enell, M. & J. Henriksson-Fejes, 1992. Ringsjön – ekosystem i förändring 1991 - Beskrivning av sjöns struktur och funktion 1991 samt förändringar under 1975–1991 (Lake Ringsjön – Ecosystem under change 1991 – Description of the structure and function of Lake Ringsjön 1991 and changes during 1975–1991). Mellanskånes Kommunalförbund and Institute of Water and Air research (IVL), 125 pp. (In Swedish).

Enell, M., J. Fejes & A. Henriksson, 1993. Ringsjön – ekosystem i förändring 1992 – Beskrivning av sjöns struktur och funktion 1992 samt förändringar under 1975–1992 (Lake Ringsjön – Ecosystem under change 1992 – Description of the structure and function of Lake Ringsjön 1992 and changes during 1975–1992). Mellanskånes Kommunalförbund and Institute of Water and Air research (IVL), 111 pp. (In Swedish).

Enell, M., J. Fejes & A. Henriksson, 1994. Ringsjön – ekosystem i förändring 1991–1993 – Beskrivning av sjöns struktur och funktion 1991–1993 samt förändringar under 1975-1993 (Lake Ringsjön – Ecosystem under change 1991–1993 – Description of the structure and function of Lake Ringsjön 1991–1993 and changes during 1975–1993). Mellanskånes Kommunalförbund and Institute of Water and Air research (IVL), 152 pp. (In Swedish).

Enell, M. & J. Fejes, 1995. Ringsjön – ekosystem i förändring 1994 – Beskrivning av sjöns struktur och funktion 1994 samt förändringar under 1975–1994 (Lake Ringsjön – Ecosystem under change 1994 – Description of the structure and function of Lake Ringsjön 1994 and changes during 1975-1994). Mellanskånes Kommunalförbund and Institute of Water and Air research (IVL), 112 pp. (In Swedish).

Enell, M., J. Fejes & C. Williams, 1996. Ringsjön – ekosystem i förändring 1995 – Beskrivning av sjöns struktur och funktion 1995 samt förändringar under 1975–1995 (Lake Ringsjön – Ecosystem under change 1995 – Description of the structure and function of Lake Ringsjön 1995 and changes during 1975–1995). Mellanskånes Kommunalförbund and Institute of Water and Air research (IVL), 86 pp. (In Swedish).

Enell, M., J. Fejes & C. Williams, 1997. Ringsjön – ekosystem i förändring 1996 – Beskrivning av sjöns struktur och funktion 1996 samt förändringar under 1975–1996 (Lake Ringsjön – Ecosystem under change 1996 – Description of the structure and function of Lake Ringsjön 1996 and changes during 1975–1996). Mellanskånes Kommunalförbund and Institute of Water and Air research (IVL), 86 pp. (In Swedish).

Enell, M. & S. Löfgren, 1988. Phosphorus in interstitial water: method and dynamics. Hydrobiologia 170: 103–132.

Granéli, W. 1979. The influence of *Chironomus plumosus* larvae on the exchange of dissolved substances between sediment and water. Hydrobiologia 66: 149–159.

Granéli, W.& D. Solander, 1988. Influence of aquatic macrophytes on phosphorus cycling in lakes. Hydrobiologia 170: 245–266.

Hamrin, S. F., E. Bergstrand, G. Cronberg, A. Persson, P. Romare & J. Strand, 1993. Sjörestaurering genom cyprinidreduktion – Ringsjöns utveckling under perioden 1987 – 1992 (Lake rehabilitation through cyprinid reduction – The development of Lake Ringsjön 1987 – 1992). Dept. of Limnology, Lund University and Swedish Fisheries Research Board, Drottningholm, 28pp. (In Swedish).

Hamrin, S. F., T. Soler, M. Eriksson, J. Svensson, H. Linge, G. Cronberg & P. Romare, 1998. Från sediment till fisk – en översiktlig studie av Vombsjöns ekosystem 1994–95 (From sediment to fish – an outline of the Lake Vombsjön ecosystem 1994–95). Fiskeriverket Rapport (1998) 1: 77–105. (In Swedish).

Hamrin, S., 1999. Planning and execution of the fish reduction in Lake Ringsjön. Hydrobiologia 404: 59–63.

Hansson, L.-A., M. Enell & E. Bergman, 1999. Lake Ringsjön: its catchment area, its history and its importance. Hydrobiologia 404: 1–7.

Hieltjes, A. H. M. & L. Lijklema, 1980. Fractionation of inorganic phosphates in calcareous sediments. J. Envir. Qual. 9: 405–407.

Holtan, H., L. Kamp-Nielsen & A. O. Stuanes, 1988. Phosphorus in soil, water and sediment: an overview. Hydrobiologia 170: 19–34.

Kairesalo, T., L. Tuominen, H. Hartikainen & K. Rankinen, 1995. The role of bacteria in the nutrient exchange between sediment and water in a flow-through system. Microb. Ecol. 29: 129–144.

Kamp-Nielsen, L., 1997. Nutrient dynamics and modelling in lakes and coastal waters. In K. Sand-Jensen & O. Pedersen (eds), Freshwater Biology – Priorities and Development in Danish Research, G. E. C. Gad Publishers, Copenhagen: 116–137.

Pettersson, K., B. Boström & O.-S. Jacobsen, 1988. Phosphorus in sediments-speciation and analysis. Hydrobiologia 170: 91–101.

Phillips, G., R. Jackson, C. Bennett & A. Chilvers, 1994. The importance of sediment phosphorus release in the restoration of very shallow lakes (The Norfolk Broads, England) and implications for biomanipulation. Hydrobiologia 275/276: 445–456.

Ryding, S.-O., 1983a. Ringsjöområdet – ekosystem i förändring (Lake Ringsjön – ecosystem under change). Mellanskånes Planeringskommitté and Dept. Of Limnology, Uppsala University, 159 pp. (In Swedish).

Ryding, S.-O., 1983b. Vattenkvalitet och ämnestransport – Ringsjön och dess tillflöden 1975–1980 (Water quality and solute transport – Lake Ringsjön and its tributaries 1975–1980). Mellanskånes Planeringskommitté and Dept. Of Limnology, Uppsala University, 82 pp. (In Swedish).

Schöllhorn, E. & W. Granéli, 1997. The importance of increasing N/P ratios for phytoplankton biomass and species composition in an oligotrophic lake. Verh. int. Ver. Limnol. 26: 615–620.

Svensson, J. M., E. Bergman & G. Andersson, 1999. Impact of cyprinid reduction on the benthic invertebrate community and implications for increased nitrogen retention. Hydrobiologia 404: 99–112.

Verner, B., 1994. Control of external loading – Lake Ringsjön, Sweden. In M. Eiseltová (ed.), Restoration of Lake Ecosystems - a Holistic Approach, IWRB Publication 32: 103–108.

Hydrobiologia **404:** 27–40, 1999.
L.-A. Hansson & E. Bergman (eds), Nutrient Reduction and Biomanipulation as Tools to Improve Water Quality: The Lake Ringsjön Story.

Qualitative and quantitative investigations of phytoplankton in Lake Ringsjön, Scania, Sweden

Gertrud Cronberg
Department of Limnology, Insitute of Ecology, Ecology Building, University of Lund, SE-223 62 Lund, Sweden

Key words: phytoplankton, fish reduction, phosphorus, biomanipulation, blue-green algae, cyanobacteria, *Microcystis*, *Anabaena*, *Aphanizomenon*, *Aulacoseira*, cryptomonads

Abstract

The algal record from Lake Ringsjön covers a period of more than 100 years. Towards the end of the 19th century, the lake had a well-functioning commercial fishery, when a brownish mess began to appear in the water, clogging-up the fishing nets and making fishing difficult. This was the first record of algal problems in the lake. Following complaints from the fishermen, the algal flora was investigated and the mess was stated to be a mass development of diatoms belonging to the genus *Melosira* (*Aulacoseira*). Diatom maxima then appeared regularly in spring and autumn, whereas blue-green algae only occurred occasionally and in low abundances. Between 1900 and 1950, nutrient concentrations slowly increased in the lake. Algal blooms of *Anabaena lemmermannii* and *Gloeotrichia echinulata* began to appear in summer. Between 1960 and 1980, the lake developed into a hypertrophic status with extensive blooms of blue-green algae from May to October, including high biomasses of mainly *Microcystis* spp. This severe pollution started with increased usage of the lake, increased tourism and recreation, intensified farming with the introduction of artificial fertilization, and also the diversion of sewage water from a sewage treatment plant. In 1968, cattle death was reported and in 1981, algal toxicity was verified by mouse bioassay. After a period of extremely poor water quality, the lake began to show signs of improvement as a result of the nutrient and fish reduction programmes. The very heavy blooms of blue-green algae decreased and the duration of blooms became shorter. The *Microcystis* species were to a certain extent replaced by *Anabaena* and *Aphanizomenon* species, and biodiversity of algae increased. With respect to the algal community, the lake has improved considerably, but is still eutrophic and suffers summer blooms of toxic blue-green algae.

Introduction

The phytoplankton flora of Lake Ringsjön was first investigated more than 100 years ago after complaints of a mess clogging the fishermen's nets (Trybom, 1893). The mess was a mass development of diatoms belonging to the genus *Melosira* (*Aulacoseira*). Since then, the plankton of Lake Ringsjön has been subjected to several investigations (Trybom, 1888; Lemmermann, 1904; Cleve-Euler, 1938; Andersson, 1948; Lundh, 1951; Björk & Lettervall, 1968; Skulberg, 1979; Cronberg, 1983). The progression of the lake from mesotrophic to hypertrophic status was made obvious by these investigations. Moreover, since my family moved to a summer house close to Lake Ringsjön, I have myself been able to follow the changes in the lake from 1946 up till now. In the middle of the 1950s, algal blooms appeared, and after swimming, several members of the family came down with itches and skin rash, probably caused by toxic *Gloeothrichia echinulata*. We ate crayfish from the lake, but by the end of the 1950s they had disappeared. In the 1970s we stopped swimming because of the mass development of stinking blue-green algae.

This review of the phytoplankton is particularly valuable as it could be verified by re-investigation of earlier collected plankton material. At the department of limnology, we have access to a large plankton archive with collections from different periods and places. The Swedish algologist E. Teiling donated his plankton collection to the department (comprising samples from about 1920 to 1960), including samples from Lake Ringsjön. Thus, as a complement to recent samples, I have used old and well preserved samples from the lake to achieve a correct, and agreed upon, identification of the taxa.

The aim of my paper is to present a review of the phytoplankton development in Lake Ringsjön during the last 100 years. Specific emphasis has been devoted to the period 1984–1996, which partly constitutes the lake's worst period, as well as the start of the recovery.

Study site

Lake Ringsjön is situated in southernmost Sweden (55° 52′ N; 13° 32′ O) and consists of three basins: Sätofta Basin (4.2 km^2, mean depth 3.0 m, maximum depth 17.0 m), Eastern Basin (20.5 km^2, mean depth 6.1 m, maximum depth 16.4 m) and Western Basin (14.8 km^2, mean depth 3.1 m, maximum depth 5.4 m) (Hansson et al., 1999). As is the case with many other lakes in urban and agricultural areas, Lake Ringsjön became eutrophicated during the 1960s and 1970s. In the beginning of the 1980s, a *nutrient reduction programme* was initiated, which reduced the external input of nutrients to the lake from phosphorus loadings exceeding 30 tons per year to around 10 tons per year. About ten years later (1989–1990), a *cyprinid fish reduction programme* (biomanipulation) was carried out in Sätofta Basin removing about 100 tons of fish (about 60% of the standing stock). During the winter of 1988, there was a fish-kill in Eastern Basin, eliminating about 500 tons of fish (about 80%) (Hamrin et al., 1999). The biomanipulation in Western Basin, which was performed in 1992, removed about 120 tons of fish (about 50%).

Methods

Sampling

From the end of 1800 to around 1960, the plankton in Lake Ringsjön was collected with a plankton net, Müller silk bolting cloth No 25, with about 55–60 μm meshes. Thus, only large-sized phytoplankton was retained and small forms were lost. The samples were mostly preserved with a mixture of alcohol and formalin.

The first quantitative investigation of phytoplankton from Lake Ringsjön was carried out in 1966–1967. Samples were collected monthly at different depths in the centre of the three main lake basins. The quantitative samples were preserved with Lugol's solution, and the 45 μm net samples with formalin to 4% final solution.

During the period 1975–1980, only quantitative samples were collected. The samples were taken with a 2 m tube sampler from surface to 2 m depth from different sampling locations in the center of the three basins. The samples were collected once a week from April to October, and were preserved with Lugol's solution. However, the samples from the different months were pooled together and evaluted as monthly mean samples.

From 1981 to the present date both quantitative and qualitative phytoplankton samples were collected monthly. During 1981–1982, the quantitative samples were taken at different depths above the deepest parts of the basins. In addition, samples were collected with a 2 m tube sampler (0–2 m depth) at several sites in the centers of each basin. The samples were pooled together in a bucket and samples for phytoplankton biomass, chlorophyll *a* and water chemistry were taken from these integrated samples. The quantitative phytoplankton samples were preserved with 8 drops of Lugol's solution to 100 ml water sample. Qualitative phytoplankton samples were collected with 10 and 45 μm plankton nets and preserved with formalin.

From 1984 to 1996, the quantitative phytoplankton samples were only taken as integrated samples as described above. Two quantitative samples were collected and one was preserved with Lugol's solution, and one with formalin to 1% final solution. The quantitative formalin preserved samples were used for counting picoplankton. Net samples were gathered with 10 and 45 μm mesh width and preserved with formalin to 4% final solution.

Quantitative phytoplankton analyses

Light microscopy

Phytoplankton was counted in sedimentation chambers (2.5 and 10 ml) with an inverted Nikon phase contrast microscope at 100×, 200× and 400× magnification. Sedimentation time was 4 to 12 h depending on the size of chambers. Single cells were counted and filamentous algae measured with a special eyepiece graticule, where a grid divided in equal sized squares, covered the whole field of view. In order to obtain acceptable reliability in estimations, 60–100 cells were counted. The filamentous algae were measured with the square grid in the eyepiece and total length of filaments (e.g. *Aulacoseira, Aphanizomenon, Planktothrix*) per diagonal of chamber or field of view were calculated. Colony-forming blue-green algae were mostly counted as colonies, (biometric determination of colony volume were attempted) and multiplied with estimated cells per colony. Samples with very dense blooms of *Microcystis, Anabaena* and *Aphanizomenon* were sonicated before being counted. The sonicated samples were then settled in 2 ml chambers and single cells were counted or total length of small filaments measured (Cronberg, 1982).

Epiflouresence microscopy

Periodically, large amounts of minute algal cells appeared, mostly colonial small blue-green algae (pico-blues). In order to be able to count these minute, yet frequently occurring, blue-green algal cells, formalin preserved quantitative samples were used. The samples were sonicated, then stained with acridinorange and counted in a Zeiss epiflourescence microscope (Cronberg, 1982). The pico-blues were divided into collective groups according to size and form. When colony-forming pico-blues were abundant, the samples were settled for at least 24 h in 2 ml sedimentation chambers. The pico-blues were counted as colonies, and the number of cells per colony were estimated. In that way, the different colony-forming pico-blues could be separated into species. The measurements of the pico-blues were made at 1025 × magnification with a normal light microscope.

Qualitative investigation of phytoplankton

The qualitative investigation of net samples were performed with a Zeiss microscope at different magnifications. The phytoplankton species were identified and their abundance noted. The Department of Limnology in Lund has a large plankton archive with preserved samples from different lakes in southern Sweden, collected from the end of 1800 up to the present date. Phytoplankton samples from Eastern Basin and Western Basin are included in the archive and old yet well preserved samples from 1896 were re-investigated. The samples studied from the archive were collected in 1896, 1900, 1915, 1949 and 1952–54. Old algal names were replaced by modern ones. In spite of the ongoing debate we will use the term 'blue-green algae' instead of 'cyanobacteria' or 'cyanoprocaryota' throughout this book.

Results 1885–1893

As early as 1892, a bloom of the diatom *Melosira* spp. (*Aulacoseira*) occurred in Lake Ringsjön. The turbidity of this bloom continued during the whole of winter and not before May 1893 the water became clear with a high transparency. In 1885 and 1887, Trybom (1888) carried out an investigation of plankton and sediment in Eastern and Sätofta Basins. The most common blue-green algae were found to be *Gloeotrichia echinulata, Merismopedia* and *Nostoc*. Frequently occurring green algae were the filamentous *Hyalotheca, Oedogonium* and *Zygnema*, the chlorococcal green algae belonging to the genera *Pediastrum* and *Scenedesmus*. In addition, the diatoms *Melosira* spp. were very common. However, by the end of the 19th century, no plankton blooms of blue-green algae were reported.

1896–1900

Samples collected during 1896–1900 from Eastern and Western Basins were re-investigated. These were probably the same samples sent to Germany and studied by Lemmermann (1904). Five samples collected between August and September were studied. The results were published together with other plankton investigations of Swedish lakes (op. cit.). The plankton communities of the two basins were similar. Common blue-green algae were *Anabaena lemmermannii, Microcystis aeruginosa* and *Woronichinia naegeliana*. The diatom *Fragilaria crotonensis* mass developed in Western Basin and other common diatoms were *Asterionella formosa* and *Acanthoceras zachariasii*. The dinoflagellate *Ceratium hiriudinella* was frequently occurring in Eastern Basin. Lemmermann recorded 71 algal species and considered Lake Ringsjön as mesotrophic. Increased concentrations of organic and inorganic nutrients caused occasional mass developments of blue-green algae, diatoms or dinaflagellates.

1938

Cleve-Euler (1938) investigated the *Melosira* plankton from different lakes in Sweden. Lake Ringsjön was included in this study, however no date was given for investigated samples. *Melosira islandica* and *M. varians* were found in large numbers. The species *M. granulata* was not recorded, despite it being common at the beginning of the century (Lemmermann, 1904).

Table 1. General observations and dominant phytoplankton species in Western and Eastern Basins of Lake Ringsjön in 1947, before the eutrophication period. Based on data from Andersson (1948). Sampling stations were the outflows of Western and Eastern Basins, respectively

	Western basin	Eastern Basin
January–April	Phytoplankton scarce Zooplankton numerous	Phytoplankton scarce Zooplankton numerous
May	*Dinobryon* cf. *bavaricum*	
June	*Ceratium hirundinella* *Gloeotrichia echinulata*	*Anabaena flos-aquae* *Gloeotrichia echinulata*
July	*Melosira islandica*	*Asterionella formosa*
August	*Melosira islandica*	*Microcystis aeruginosa*
September	*Melosira islandica*	*Microcystis aeruginosa*
October	*Melosira islandica* (mass development)	*Melosira* sp.

1947

In 1947, Andersson (1948) made the first yearly investigation of the plankton development in Western Basin (Table 1). In July–October 1947, the western part of Western Basin had rich development of the diatom *Melosira islandica*, while the eastern part was dominated by the blue-green algae *Microcystis aeruginosa* (Andersson, 1948). This indicates that Eastern Basin was probably even more richer in nutrients than Western Basin, and that the eutrophication of Lake Ringsjön started earlier in Eastern Basin than in Western Basin. However, the lake had a relatively low nutrient concentration, since the yearly mean Secchi depth was 3.2–3.6 m in the two sampling stations of Western Basin.

1949

In 1949, Lundh (1951) investigated phytoplankton from Eastern Basin and Western Basin. In April, Western Basin was dominated by the chrysophyte *Dinobryon* and in Eastern Basin the diatom *Asterionella formosa.* In September, the diatom *Fragilaria crotonensis* was most common in Western Basin and the dinoflagellate *Ceratium hirundinella* in Eastern Basin. About 60 phytoplankton species were recorded and about 30% consisted of blue-green algae. Lundh (1951) incorporated Lake Ringsjön among the eutrophic lakes, but Lake Ringsjön did not belong to the Myxophyceae-rich lakes (blue-green algae in nutrient-rich lakes).

1966–1967

Because of this, the drinking water supplies for the growing cities in Scania had to be secured and limnological investigations were carried out in Lake Ringsjön in 1966 and 1967 (Björk & Lettervall, 1968). The investigations also included a study of the phytoplankton development. For a year and a half, monthly samples were taken in all three basins at different depths. The biomass of phytoplankton varied within the different basins. In July 1967, the highest biomass, 43 mg l^{-1}, was recorded in Eastern Basin (Figure 1). Sätofta and Western Basins developed maxima in August 1967. At all three occasions, diatoms were dominant, mainly *Asterionella formosa, Cyclotella, Stephanodiscus, Melosira* and *Fragilaria crotonensis.* Only in June 1967 in Eastern Basin, the blue-green algae *Anabaena* and *Aphanizomenon* were frequent. Thus, Lake Ringsjön was still dominated by diatoms during the summers and the blue-green algae were of minor importance.

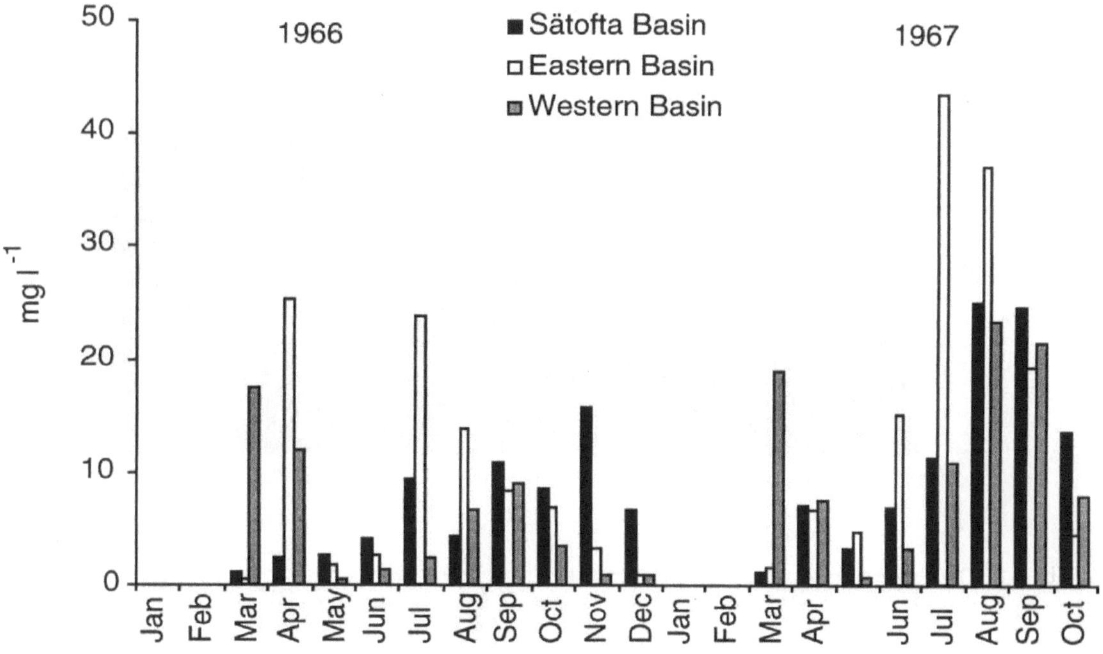

Figure 1. Biomass of phytoplankton (mg l^{-1}) in Eastern, Western and Sätofta Basins of Lake Ringsjön during 1966 and 1967. Data from Björk & Lettervall (1968).

Archive samples 1896–1980

In order to get a more thorough picture of the phytoplankton development in Lake Ringsjön, I analyzed net samples from the plankton archive at the department of Limnology in Lund (see above). The samples were generally collected between June and September. About 70 algal species were identified. From the end of 1800 to 1960s, about 25–50% consisted of blue-green algal species, and the rest was mainly diatoms and green algae. Periodically, dinoflagellates and chrysophytes were numerous. From the beginning of the 1970s the plankton community began to change. The blue-green algal species occurred more frequently around 1980 and more than 75% of recorded species were blue-greens. Dinoflagellates and chrysophytes disappeared completely in the summers, but could be found in spring or autumn. The number of diatoms and green algae were reduced (Figures 2a–b).

Changes started in Eastern Basin first and somewhat later in Western Basin. There were very few investigations in Sätofta Basin, therefore the algal development of this basin was not known until 1966. During the 1940s, nitrogen-fixing algae like *Anabaena, Aphanizomenon* and *Gloeotrichia*, became common and reached maximum development in the early 1970s. In the middle of the 1970s, however, nitrogen-fixing blue-green algae decreased and non-fixing blue-green algae, such as *Microcystis* spp., became dominant and formed blooms lasting from June to October.

Some algal species characteristic for Lake Ringsjön were common at the end of the 19th century and up till 1970, then disappeared completely, such as *Gloeothrichia echinulata, Acanthoceras zachariasii, Rhizosolenia longiseta, Ceratium hirundinella, Melosira varians, Tabellaria fenestrata* var. *asterionelloides, Staurastrum* spp., *Staurodesmus* spp. and *Spondylosium planum*. Other species were greatly reduced, namely *Anabaena circinalis, A. lemmermannii, Dinobryon* spp., *Asterionella formosa, Fragilaria crotonensis* and *Cyclotella* spp.

During 1970–1985, the blue-green algae increased considerably in abundance and dominated completely in the summers. In 1960–1975, *Aphanizomenon flos-aquae, Anabaena crassa, A. spiroides* and *A. flos-aquae* were most frequent, but in 1975–1985 they were replaced by *Microcystis aeruginosa, M. flos-aquae, M. viridis* and *M. wesenbergii*.

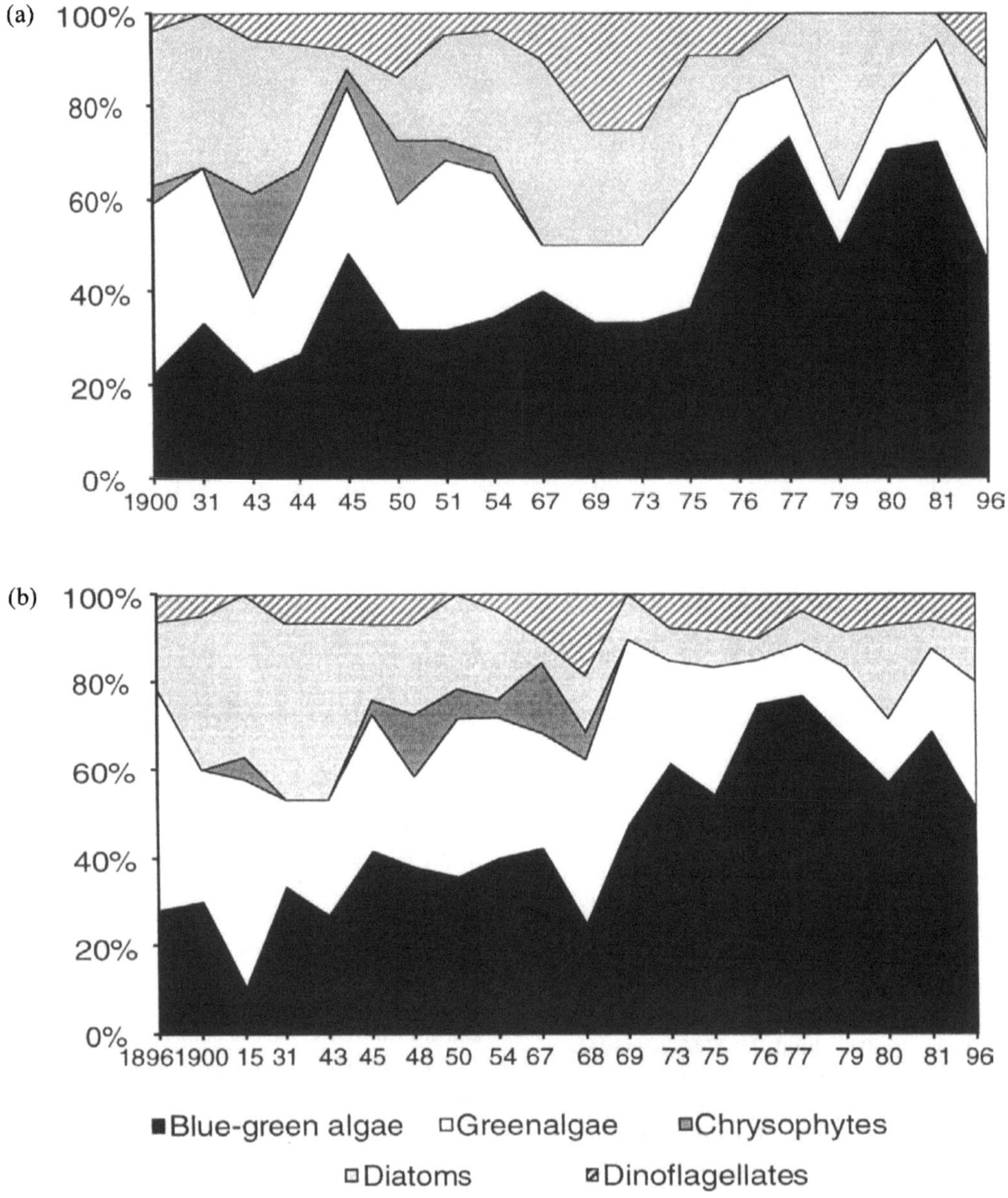

Figure 2. Percentage distribution of blue-green algae, green algae, chrysophytes, diatoms and dinoflagellates in (a) Eastern Basin and (b) Western Basin during 100 years, 1896–1996.

1984–1996

Monthly investigations of phytoplankton were carried out during 1981–1982 and from 1984 to date. Here, the period 1984–1996 will be reviewed in detail.

Phosphorus

From 1984 to 1986 during January–May, the monthly mean inflow of phosphorus was high (600–1 400 kg tot-P month^{-1}), but decreased between 1986 and 1993 (300–600 kg tot-P month^{-1}; Figure 3). In 1994–1996 however, the inflow of phosphorus increased

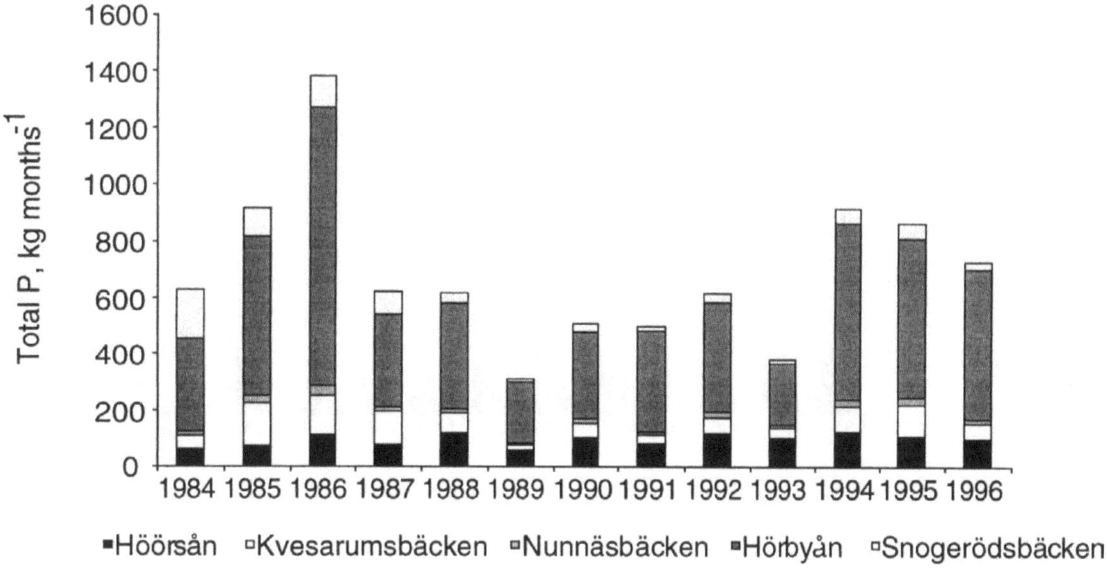

Figure 3. The total load of phosphorus (monthly mean values from January to May) to Lake Ringsjön from the different tributaries, 1984-1996.

again (700–900 kg tot-P month^{-1}). The change in phosphorus load may have been affected by climatic factors, such as lesser rainfall in the spring, but also by active efforts at reducing the phosphorus load to the lake by reduced nutrient input from surrounding farms and urban areas (Figure 3). In spite of this, the algal blooms showed no signs of diminishing.

Biomass of phytoplankton

In the beginning of the 1980s, the highest biomass of phytoplankton was recorded in Eastern Basin, more than 100 mg l^{-1}. In Sätofta Basin, up to 70 mg l^{-1} was measured, and finally in Western Basin, the maximum was about 50 mg l^{-1}. Western Basin generally had lesser water blooms than the two basins upstream. In 1988, the distinct decrease in phosphorus and nitrogen to the lake, plus the fish-kill in Eastern Basin, might have led to the reductions in algal biomass in all three basins. The largest decrease was observed in Eastern Basin. The reduction in algal biomass was mainly a decrease in the blue-green algae *Microcystis* spp. (Figures 4 & 5).

Below, I will focus on the fluctuations in the phytoplankton community of Eastern Basin, which lies between the two others. In all basins however, similar changes were observed in both biomass and species composition of phytoplankton.

Blue-greens

During the period of 1984–1996, the largest changes in phytoplankton were recorded within the blue-green algal community. From the 1970s to 1988, the genus *Microcystis* dominated. Most common were the species *Microcystis aeruginosa, M. flos-aquae, M. viridis* and *M. wesenbergii*. In 1988, these algae were much reduced. They still appeared in the succeeding summers, but with low biomass and for short periods (Figure 5). Instead, the nitrogen fixing filamentous blue-green algae increased. In the summers of 1989 and 1991, species belonging to the genus *Anabaena* were especially common (Figure 5). In August 1989, *Aphanizomenon klebahnii* showed a high maximum, but appeared in succeeding summers with a low biomass (Figure 5). In August 1987, following the fish-kill, the non-nitrogen fixating blue-green algae *Planktothrix agardhii* was certainly numerous, but in the autumn of 1990, it had developed into a high maximum. In the following years, *Aphanizomenon* appeared only in small quantities. From 1987 onwards, the chroococcalean blue-greens belonging to the genera *Radiocystis, Snowella* and *Woronichinia* began to appear and are now common (Figure 6). In the period of 1984–1989 pico-blues (small blue-green algae with cell diameter 1–2 μm l^{-1}) were common, in 1990 they disappeared but reappeared in 1994. The pico-

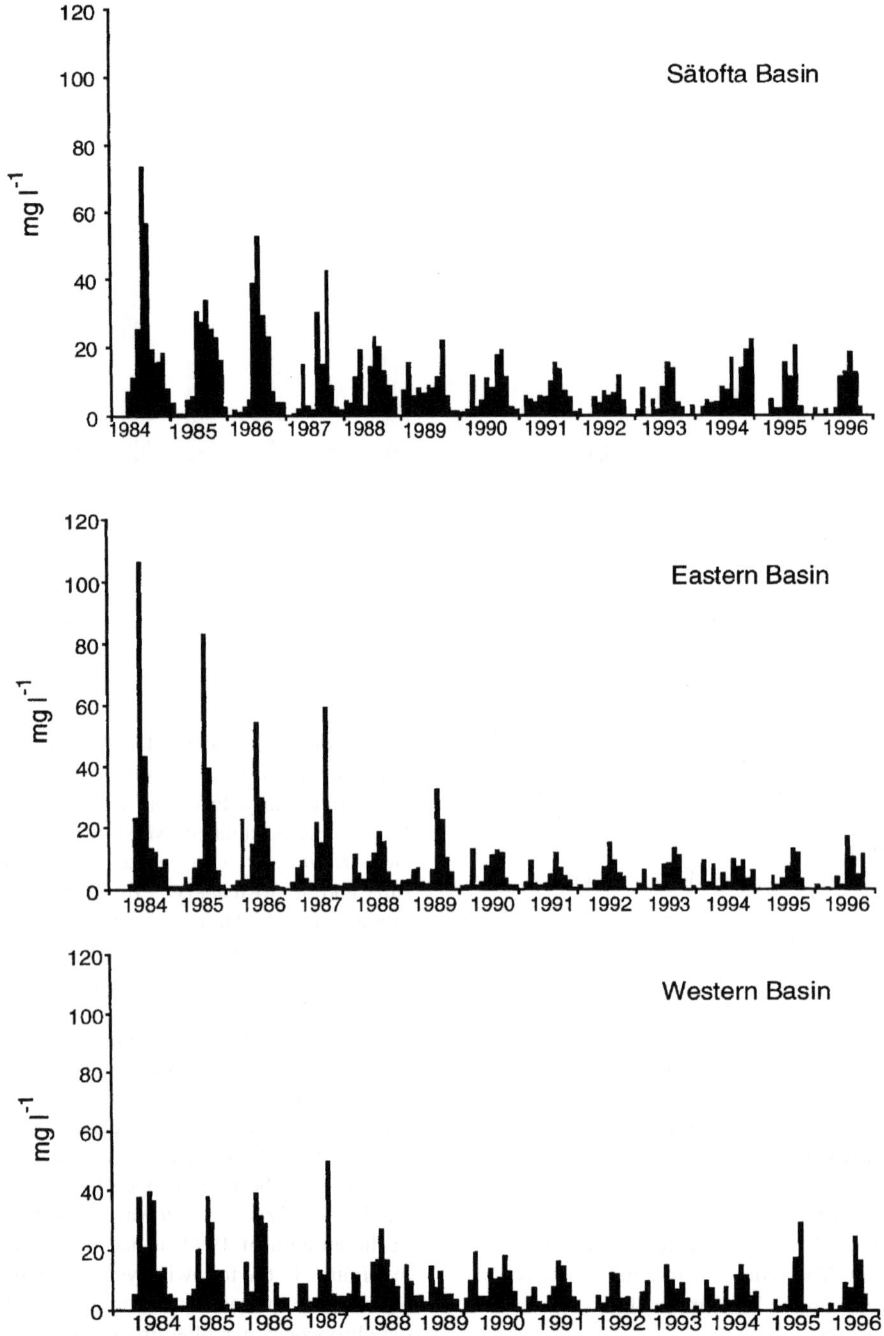

Figure 4. Monthly fluctuations in total phytoplankton biomass (mg l^{-1}) in the three basins of Lake Ringsjön from 1984 to 1996.

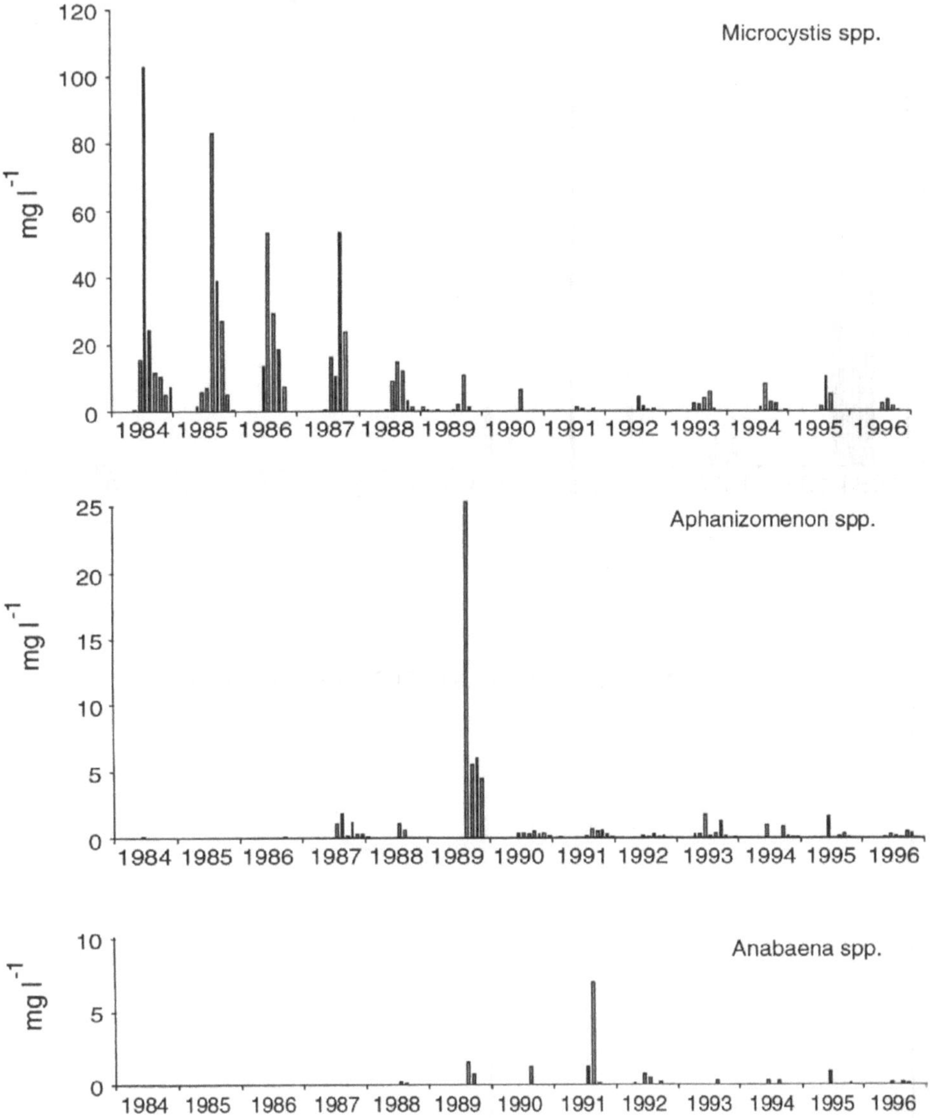

Figure 5. Monthly fluctuations in biomass (mg l^{-1}) of the blue-green algal genera *Microcystis, Aphanizomenon* and *Anabaena* in Eastern Basin from 1984 to 1996.

blues were dominated by the genera *Aphanothece, Aphanocapsa* and *Cyanodictyon* (Figure 6).

Diatoms

At the beginning of the 1980s, small amounts of the diatom genus *Aulacoseira* (=*Melosira*) were recorded. From 1986, however, high biomasses of *Aulacoseira* were recorded almost every year. The *Aulacoseira* assemblage was not identified to species level, so any changes within the genus cannot be identified. Additional common diatoms in Eastern Basin belong to the genera *Actinocyclos, Cyclotella* and *Stephanodiscus.* Within this group of diatoms, a slight increase in biomass was recorded from 1987, but decreased again in

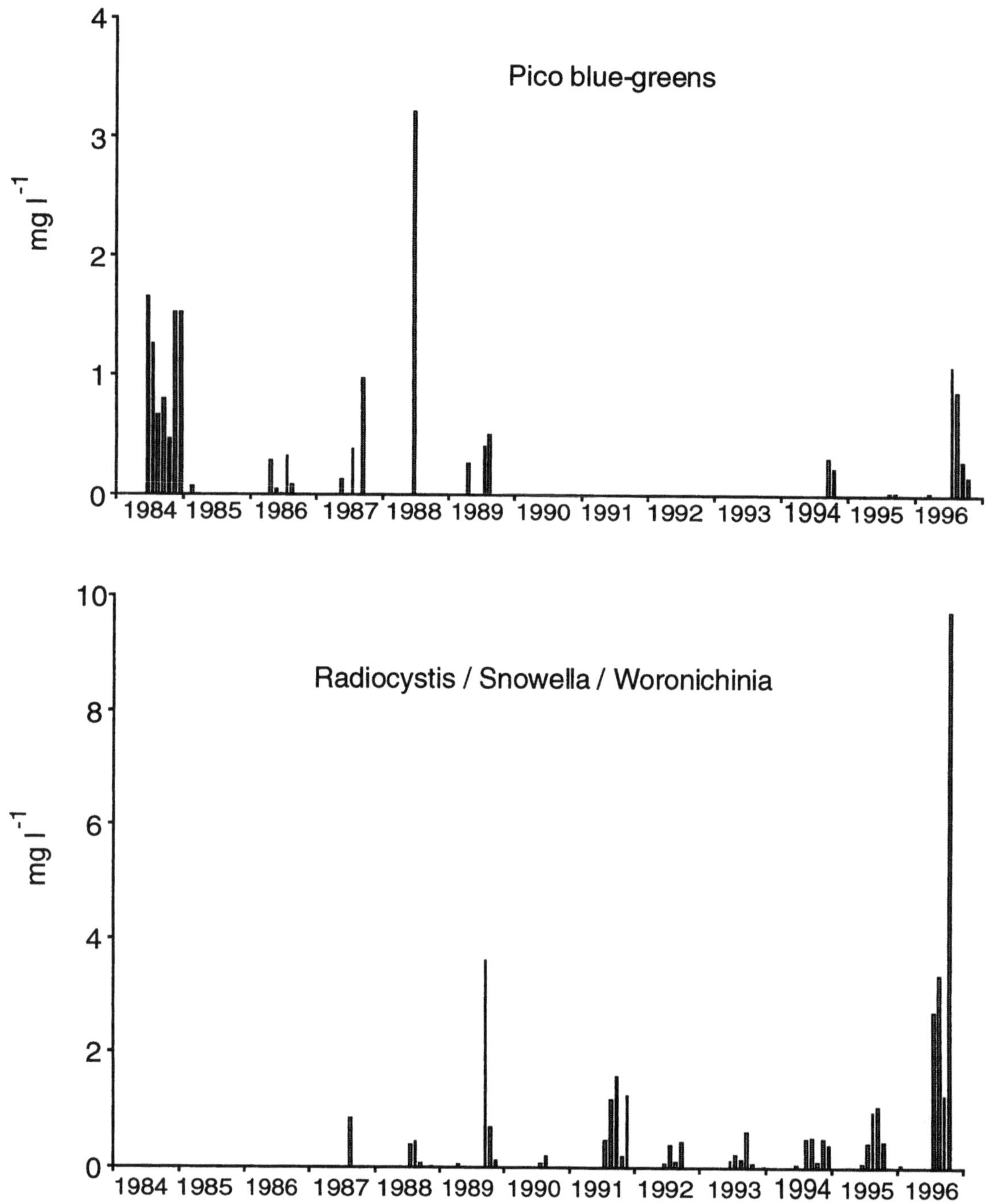

Figure 6. **Monthly fluctuations in biomass (mg l^{-1}) of the pico blue-green algae and the genera *Radiocystis, Snowella* and *Woronichinia* in Eastern Basin from 1984 to 1996.**

1993. The pennate diatoms *Diatoma, Fragilaria* and *Synedra* appeared irregularly, but were more frequent in 1987 and 1990–1992.

Dinoflagellates

In Eastern Basin, dinoflagellates were normally rare and appeared irregularly. In 1989, *Ceratium hirundinella* and *C. furcoides* were recorded for the

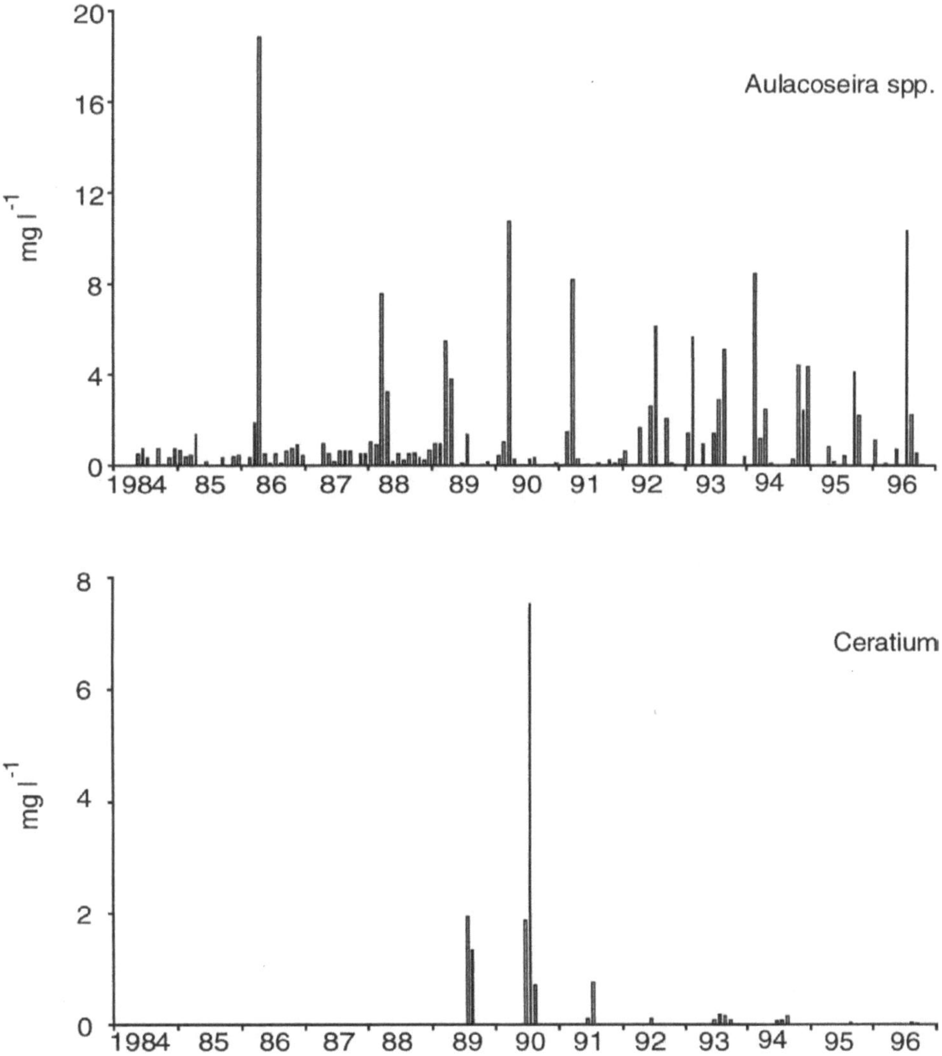

Figure 7. Monthly fluctuations in biomass (mg l^{-1}) of the diatom *Aulacoseira* spp. (former *Melosira*) and the dinoflagellate *Ceratium* spp. in Eastern Basin from 1984 to 1996.

first time, and in 1990 appeared in high abundances (7.5 mg l^{-1}). From 1991 onwards, only low numbers of *Ceratium* were recorded (Figure 7). The dinoflagellates *Enzia acuta, Gymnodinium helveticum* and *Peridinium* sp. were only recorded in 1993 and in low amounts, biomass up to 0.6 mg l^{-1}. As a whole, this group of algae has been rare in Lake Ringsjön.

Cryptomonads

The cryptomonads have always been present in Eastern Basin. Between 1987 and 1989, biomasses up to 2 mg l^{-1} were recorded. During the other years, the biomass was about 0.5 mg l^{-1} (Figure 8). The genera *Chroomonas, Cryptomonas* and *Rhodomonas* were identified.

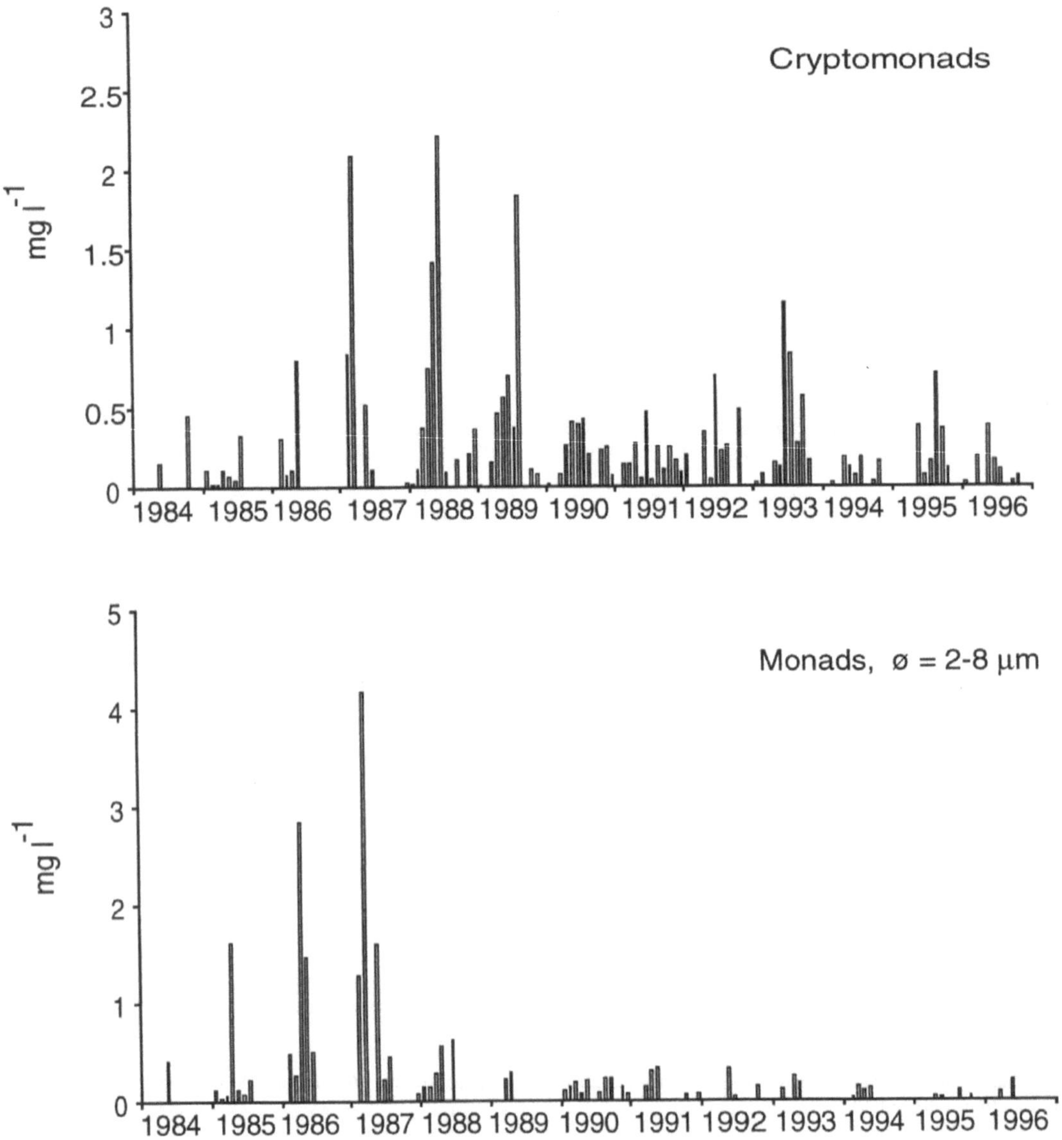

Figure 8. Monthly fluctuations in biomass (mg l^{-1}) of cryptomonads and monads (diameter 2–8 μm) in Eastern Basin from 1984 to 1996.

Monads

Monads is a collective group of unidentified algae with cell diameter between 2 and 8 μm, including rounded cells or cells with 1–2 flagella. Before 1988, the monads reached biomasses up to 4 mg l^{-1}, but from 1988 their biomass decreased to less than 0.5 mg l^{-1} (Figure 8).

Discussion

Around 1900, Lake Ringsjön was an mesotrophic lake dominated by diatoms and chrysophytes. Between 1900 and 1950, a slight change in phytoplankton biomass and composition was recorded. The lake was polluted from different sources and became ever

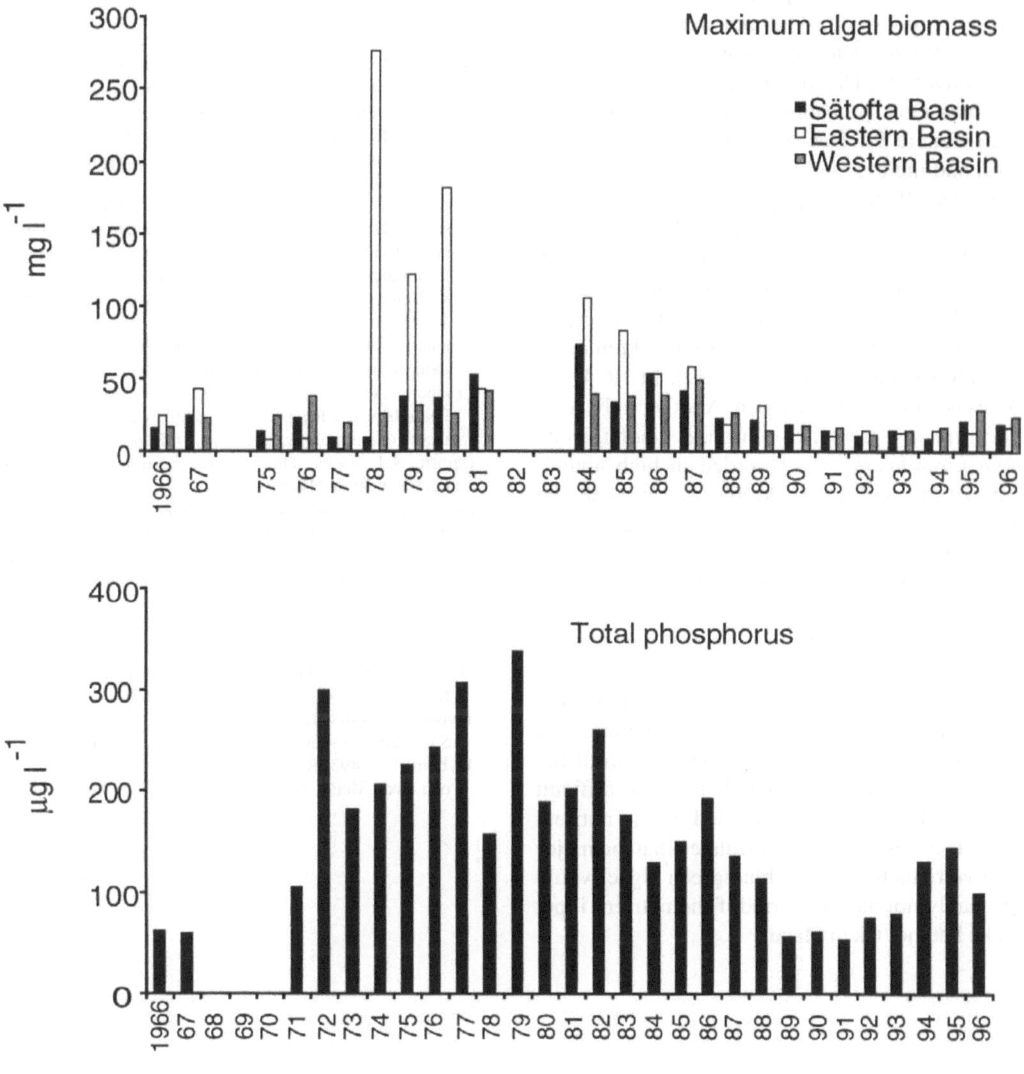

Figure 9. The maximum biomass of phytoplankton (mg l^{-1}) developed in the three basins of lake Ringsjön (upper panel), and the concentration ($\mu g\ l^{-1}$) of total phosphorus in the water (lower panel) of Eastern Basin of Lake Ringsjön from 1966 to 1996.

more nutrient rich and eutrophic. Between 1970 and 1984, the biomass production increased drastically and the lake was characterized by long-lasting, heavy blooms of blue-green algae. The lake had became hypertrophic. From 1987, the lake has successively returned to eutrophic conditions with respect to the algal community (Figures 4 & 9). Quantitative data on phytoplankton exist for 1966–1967, 1975–1981 and 1984–1996. Between 1966 and 1967 the biomass of algae in Eastern Basin reached up to about 25 mg l^{-1}. The algal biomass increased, and in 1978–1980 the maximum values were recorded, 180–276 mg l^{-1}. The biomass then began to decrease again. In 1988 it was a third of the biomass in 1984, and continued to decrease to about 10 mg l^{-1} (Figure 9). Between the years 1984 and 1996, a successive decrease in phytoplankton biomass was recorded, simultaneous to an increase in species diversity.

The most striking decrease was within the blue-green algal community. Species belonging to the genus *Microcystis* diminished and were replaced by *Anabaena, Aphanizomenon, Snowella, Radiocystis* and *Woronichinia*. The decrease in phytoplankton biomass was most probably a result of the nutrient reduction (Figure 9). However, the fish-kill in Eastern Basin and the cyprinid reduction in Sätofta Basin may also have affected the phytoplankton community. Hence, from 1988 there was a noticeable reduction in the pico-blues and for some years they disappeared. This may have been a result of increased grazing by zooplankton following the fish reductions despite the minor changes recorded in the zooplankton community (Bergman & Bergstrand, 1999). In 1994, the pico-blues began to increase again. Another indication of higher grazing pressure is that the dinoflagellate *Ceratium hirundinella* was recorded in high abundances between 1989–1991. Moreover, in 1988 the monads (sized 2–8 μm) were reduced in numbers, which may also indicate a higher number of zooplankton and more efficient grazing. During the years 1987–1989, the number of cryptomonads increased. These have the ability to multiply fast even at high grazing pressure.

Hence, despite changes in the zooplankton community following fish reductions being moderate, they may have affected the *qualitative* composition of the phytoplankton commuity. Although it may be difficult to distinguish between effects caused by the nutrient and fish reductions, it may be postulated that the major *quantitative* reduction in the blue-green algae would most probably not have occurred if the nutrient input to the lake had not taken place.

References

Andersson, A., 1948. Näringstillgång och planktonutveckling i några skånska sjöar. Vattenhygien 4: 9–21.

Bergman, E., 1999. Changes in the nutrient load and lake water chemistry in Lake Ringsjön, southern Sweden, from 1966 to 1996. Hydrobiologia 404: 9–18.

Bergman, E. & E. Bergstrand, 1999. Lack of a top-down effect on the zooplankton community after a cyprinid fish reduction. Hydrobiologia 404: 77–87.

Björk, S. & U. Lettervall, 1968. Bolmen-Lagan-Ringsjön. Resultat från en 2-årig limnologisk undersökning i samband med projektering för Skånes framtida vattenförsörjning. Sydvatten. Limnologiska Institutionen, Lund, Sweden (In Swedish).

Cleve-Euler, A., 1938. Våra sjöars Melosira-plankton. Bot. notiser 4: 143–163.

Cronberg, G., 1982. Phytoplankton changes in lake Trummen induced by restoration. Folia limnol. scand. 18: 1–119.

Cronberg, G., 1983. Ringsjöområdet ekosystem i förändring. Förändringar i Ringsjöarnas växtplankton under 1900-talet. Institute of Limnology. University of Lund, 56 pp (In Swedish).

Hamrin, S. F., 1999. Planning and execution of the fish reduction in Lake Ringsjön. Hydrobiologia 404: 59–63.

Hansson, L.-A., M. Enell & E. Bergman, 1999. Lake Ringsjön: its catchment area, its history and its importance. Hydrobiologia 404: 1–7.

Lemmermann, E., 1904. Das Plankton schwedischer Gewässer. Arkiv för Botanik 2: 1–209.

Lundh, A., 1951. Studies on the vegetation and hydrochemistry of Scanian lakes. Bot. notiser 3: 1–137.

Skulberg, O., 1979. Giftvirkning av blågrönnalger-första tillfälle av forgiftning i Norge. Temarapport 4 NIVA.

Trybom, F., 1888. Bottenprof från svenska insjöar. Geol. Fören. i Stockholm Förhandl. 10: 491–497 (In Swedish).

Trybom, F., 1893. Ringsjön i Malmöhuslän dess naturförhållande och fiske. Medd. Kongl. Lantbruksstyr. 4: 14–42 (In Swedish).

Hydrobiologia **404:** 41–51, 1999.
L.-A. Hansson & E. Bergman (eds), Nutrient Reduction and Biomanipulation as Tools to Improve Water Quality: The Lake Ringsjön Story.

Autumn waterfowl abundance in Lake Ringsjön, 1968–1996

Gunnar Andersson[1] & Leif Nilsson[2]
[1]*County Administration Board of Scania, SE-205 15, Malmö, Sweden*
[2]*Institute of Ecology, Ecology Building, SE-223 62 Lund, Sweden*

Key words: Lake Ringsjön, restoration, biomanipulation, waterfowl, bird

Abstract

Lake Ringsjön has been subject to a severe eutrophication process followed by various counter measures, such as nutrient reduction and fish manipulation. We have quantified staging waterfowl populations from 1968, before the eutrophication process reached its maximum, to 1996 when both the nutrient and fish reduction programmes were performed. The general pattern is that high numbers staged in Lake Ringsjön in 1968–1972. In the following 15 years, few waterfowl used the lake as staging area, but from about 1989, the number of birds again began to increase. The likely explanation of this pattern is an improved food availability within the lake, rather than large-scale, national, increases in bird population sizes, a notion strengthened by comparisons with standardized national waterfowl counts. We conclude that there are strong relations between limnological conditions and waterfowl abundances, and that the measures undertaken in Lake Ringsjön improved the lake as a staging area for waterfowl.

Introduction

Waterfowl are rarely included in limnological studies, although many species are dependent on food resources produced in lake ecosystems. Some of the birds are omnivorous but several species are specialists utilizing different types of food (Mitchell & Perrow, 1998). High densities of waterfowl are reported to affect abundance or biomass of their food items (Nilsson, 1972; Lauridsen et al., 1993; Mitchell & Perrow, 1998).

The first waterfowl study in Lake Ringsjön started in 1968 and focused on feeding ecology of diving ducks (Nilsson, 1972). Regular counts of most species were included in the study and continued until 1978 (Nilsson, 1974 and unpubl.). During the 1970s, abundance of waterfowl declined drastically in some shallow lakes in southern Sweden and indications were that large scale changes in lake trophic status were involved (Andersson, 1981). One of these lakes was Lake Ringsjön, and when measures to restore the lake were being planned, we realized that waterfowl abundance was a variable that may be affected. Counts were therefore recontinued in 1982, and in the present paper we present results from 1982 to 1996. Even though there are some gaps within and between the two studies, waterfowl abundances are covered for an almost 30-year period.

The main importance of Lake Ringsjön for waterfowl is as a staging area, especially during autumn migration, whereas it is of much less importance to breeding waterfowl since there are few suitable breeding islands and only small areas of reed-beds. With the exception of fish-eating birds, such as grebes, which can use the open areas of the lake hunting for pelagic fish, other species are more restricted in their use of the lake. This is mainly due to the relatively large depth, especially of Eastern Basin, whereas Western Basin is well within the diving depth of all diving ducks (Nilsson, 1972). Moreover, the shores of Lake Ringsjön are mainly stony and steep with only a narrow belt of emergent vegetation reaching a depth of about 1–2 m and offering limited feeding areas for dabbling ducks. Just a minor part of the shore is sandy and shallow, for example east of Lillö in Western Basin, at Fulltofta in Eastern Basin and parts of the shores of Sätofta Basin (Figure 1). These areas offer feeding possibilities for dabbling ducks and swans.

The objective of this paper is to evaluate the progress of the staging waterfowl community of Lake Ringsjön over time to elucidate whether changes in

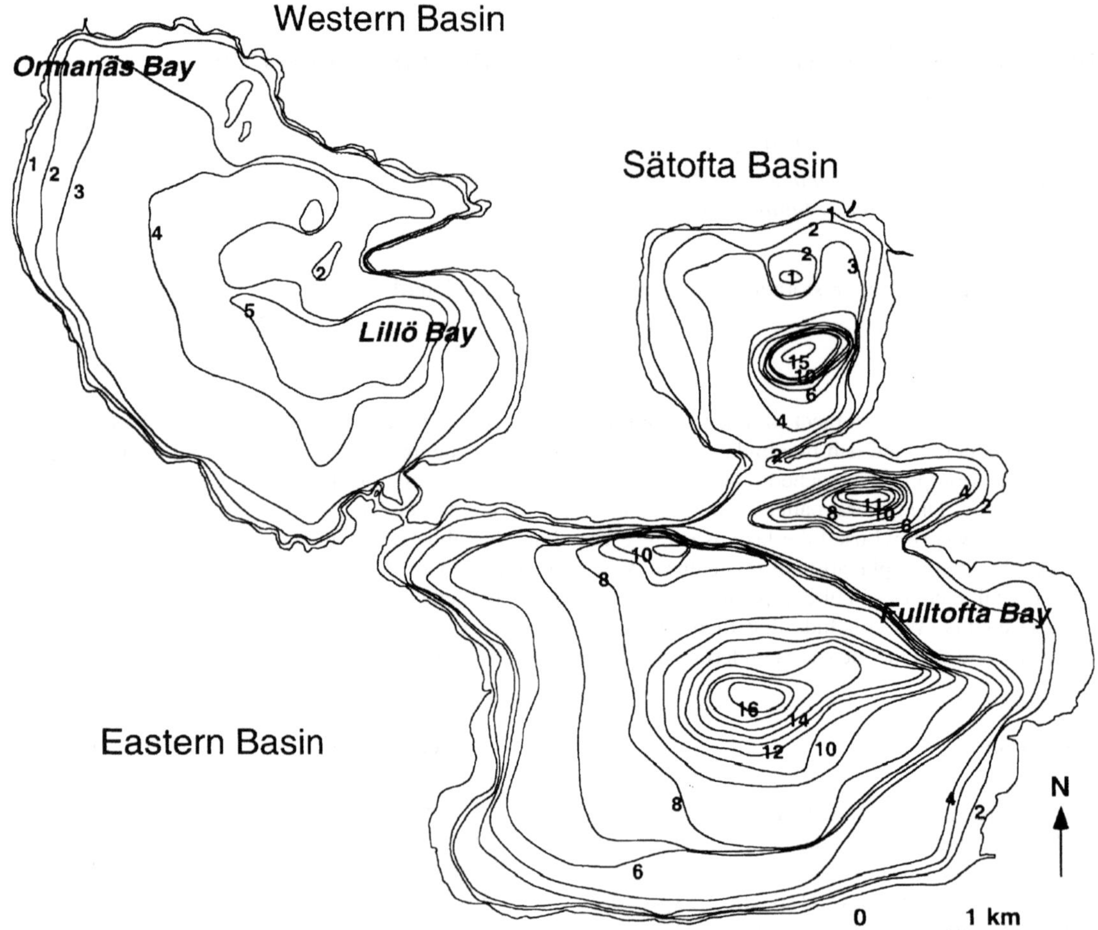

Figure 1. Map of Lake Ringsjön including Western Basin, Eastern Basin and Sätofta Basin. Depth contours (1 m intervals) and names of some important waterfowl staging areas are shown.

trophic status have had any effect on the waterfowl abundance.

Study site

Lake Ringsjön is situated in southernmost Sweden (55° 52′ N; 13° 32′ O) and consists of three basins: Sätofta Basin (4.2 km^2, mean depth 3.0 m, maximum depth 17.0 m), Eastern Basin (20.5 km^2, mean depth 6.1 m, maximum depth 16.4 m) and Western Basin (14.8 km^2, mean depth 3.1 m, maximum depth 5.4 m) (Hansson et al., 1999). As with many other lakes in urban and agricultural areas, Lake Ringsjön became eutrophicated during the 1960s and 1970s. In the beginning of the 1980s, a *nutrient reduction programme* was initiated which reduced the external input of nutrients to the lake from phosphorus loadings exceeding 30 tons per year to around 10 tons per year. About ten years later (1989–1990), a *cyprinid fish reduction programme* (biomanipulation) was carried out in Sätofta Basin removing about 100 tons of fish (about 60% of the standing stock). During the winter of 1988, there was a fish-kill in Eastern Basin, eliminating about 500 tons of fish (about 80%) (Hamrin, 1999). The biomanipulation in Western Basin, which was performed in 1992, removed about 120 tons of fish (about 50%).

Methods

From 1968/69 to 1973/74, monthly waterfowl counts were undertaken in both basins of Lake Ringsjön in August – April. In 1969, more or less weekly counts were made in late summer and autumn as a part of the special studies (Nilsson, 1972). In the years 1975–1978, counts were undertaken only once, in mid-September, whereas two complete autumn counts (in September and October) were done each year in 1982–1996. The September counts were undertaken at the same time as the National September Waterfowl Counts in Sweden (Nilsson, 1997b).

All waterfowl counts were carried out from the shore using normal binoculars and a telescope with a magnification up to 40×. The counts were made from fixed positions chosen so that the entire water surface between two positions was adequately covered. To count diving species, repeated scans over the same water areas were made with short intervals. Counts were only performed on days with favourable weather conditions, with especially the wind being of importance in this respect since excess waves could impede the detection of diving waterfowl. The counts included all waterfowl species with the exception of geese, gulls, terns and shorebirds. With few exceptions, surveys were performed by one of us; 1968–1978 (LN), and 1982–1996 (GA). The counting method was the same as that used in the International Waterfowl Counts (Nilsson, 1976, 1978). From 1984 onwards, cormorants were counted just before dawn when they were concentrated on perching sites.

One or two counts per year may seem to be too few to adequately sample changes in a waterfowl community. This may be true if the aim is to produce a detailed study of changes in numbers of waterfowl over a short period. The aim however, of the present study is to elucidate large-scale and long-term trends, where annual deviations are of less importance for the general picture. The regular counts have been performed in autumn for two reasons:

1. In September and October all birds can fly and are free to chose suitable feeding habitats;
2. Territorial behaviour is largely reduced and birds gather in flocks at suitable feeding areas.
3. Results from the National September Waterfowl Counts (Nilsson, 1997b) can be used as a reference on fluctuations in population sizes of different waterfowl species in Sweden.

Results

The results of the counts undertaken in September and October for the various waterfowl species appearing in any numbers and with any regularity at Lake Ringsjön are presented in Figures 2 – 5. The birds are not presented in systematic order but grouped according to their main food resources: herbivores, omnivores, benthivores and piscivores.

Herbivores

Both the mute swan (*Cygnus olor*) and the whooper swan (*Cygnus cygnus*) were regularly found in the lake during the first years of the survey (Figure 2). In respect to the mute swan, maximum counts in the early years was about 50 individuals. Numbers decreased and remained low until 1989, when an increase was noted. October numbers decreased as early as in 1970 although the increase from 1989 and on was not as marked as in September (Figure 2). There was not any clear tendency in September numbers of whooper swan until an increase was noted from 1989 onwards with a peak of no less than 50 individuals in September 1993. October counts of whooper swan showed another picture, with some flocks in the early years and then a break in the series of observations until 1989 and onwards, when some flocks appeared again. The whooper swans were clearly concentrated in a shallow area in the eastern part of Eastern Basin with stands of *Potamogeton pectinatus*. A few observations of the Bewick's swan (*Cygnus columbianus*) were made and these are included in the number of whooper swan.

High numbers of the herbivorous coot (*Fulica atra*) were found in 1968 and 1971 (Figure 2; coots were not counted in September 1969 and 1970 or in October 1970). Small numbers were seen in some other years in the 1970s. From 1975 to 1989, hardly any coot were found in the lake, but in the 1990s, as many as 200 individuals were again to be found in the lake in September, whereas only very few remained in October.

Omnivores

The dominant species among the dabbling ducks is the mallard (*Anas platyrhynchos*), both in September and October. Over the first years, the number of mallards was generally around 500 at the September count and at the same level in October, with the exception of higher numbers in 1971 (Figure 3). An obvious reduction in the number of mallards was noted in the

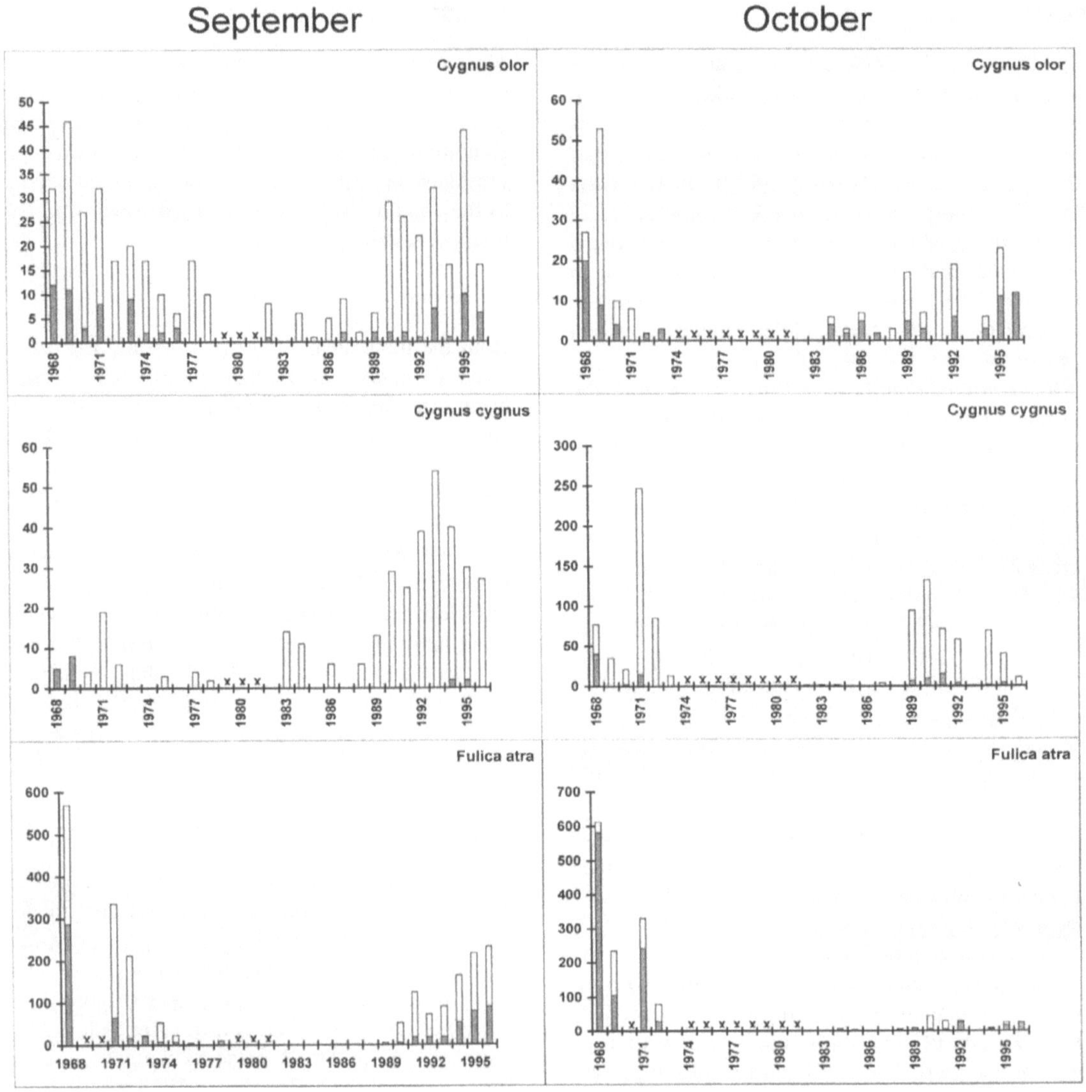

Figure 2. Number of herbivorous waterfowl counted in Western Basin (dark bar) and Eastern Basin (white bar) during September and October counts from 1968 to 1996. Herbivorous waterfowl include: mute swan (*Cygnus olor*), whooper swan (*Cygnus cygnus*) and coot (*Fulica atra*). Note the different scales. No counts were performed in September 1979–1981 or in October 1974–1981. Furthermore, coots were not counted in September 1969–1970 and in October 1970.

1980s, followed by higher numbers in 1989–1996. It may be noted in this context that there has been a shift in distribution between the two basins, with Western Basin being most important the first years and Eastern Basin in later years.

The other dabbling ducks exhibit a more irregular appearance at Lake Ringsjön (Figure 3). The teal (*Anas crecca*) and the wigeon (*Anas penelope*) were counted in high numbers most years, whereas the pintail (*Anas acuta*) was seen only on relatively few occasions. The teal numbers were low in the 1980s, as was the case with the mallard. Abundances varied widely, however, which is most probably a result of the long intervals between counts (Nilsson, 1976,

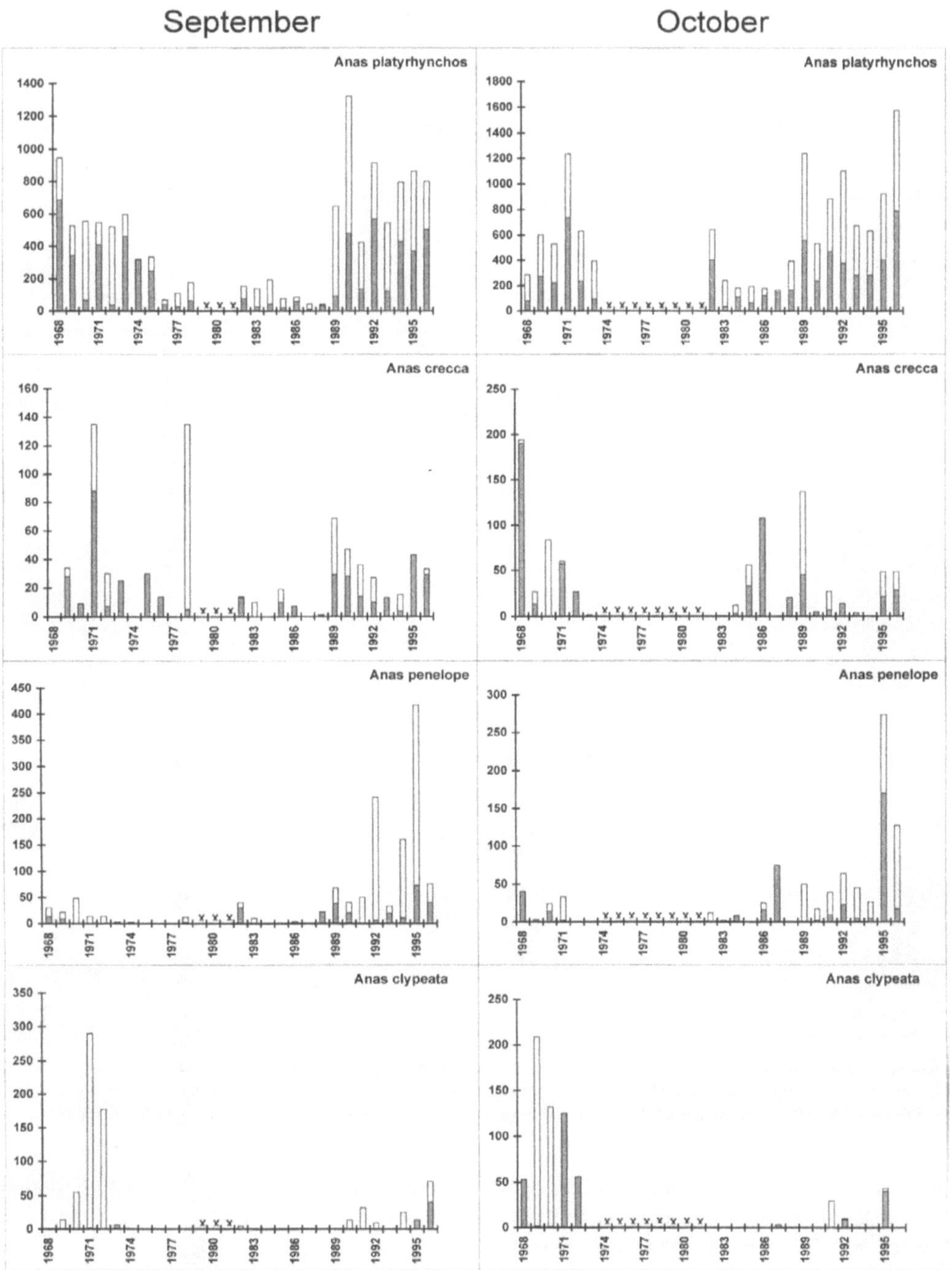

Figure 3. Number of omnivorous waterfowl counted in Western Basin (dark bar) and Eastern Basin (white bar) during September and October counts from 1968 to 1996. Omnivorous waterfowl include: the mallard (*Anas platyrhynchos*), the teal (*Anas crecca*), the wigeon (*Anas penelope*) and the shoveler (*Anas clypeata*). Note the different scales. No counts were performed in September 1979–1981 or in October 1974–1981.

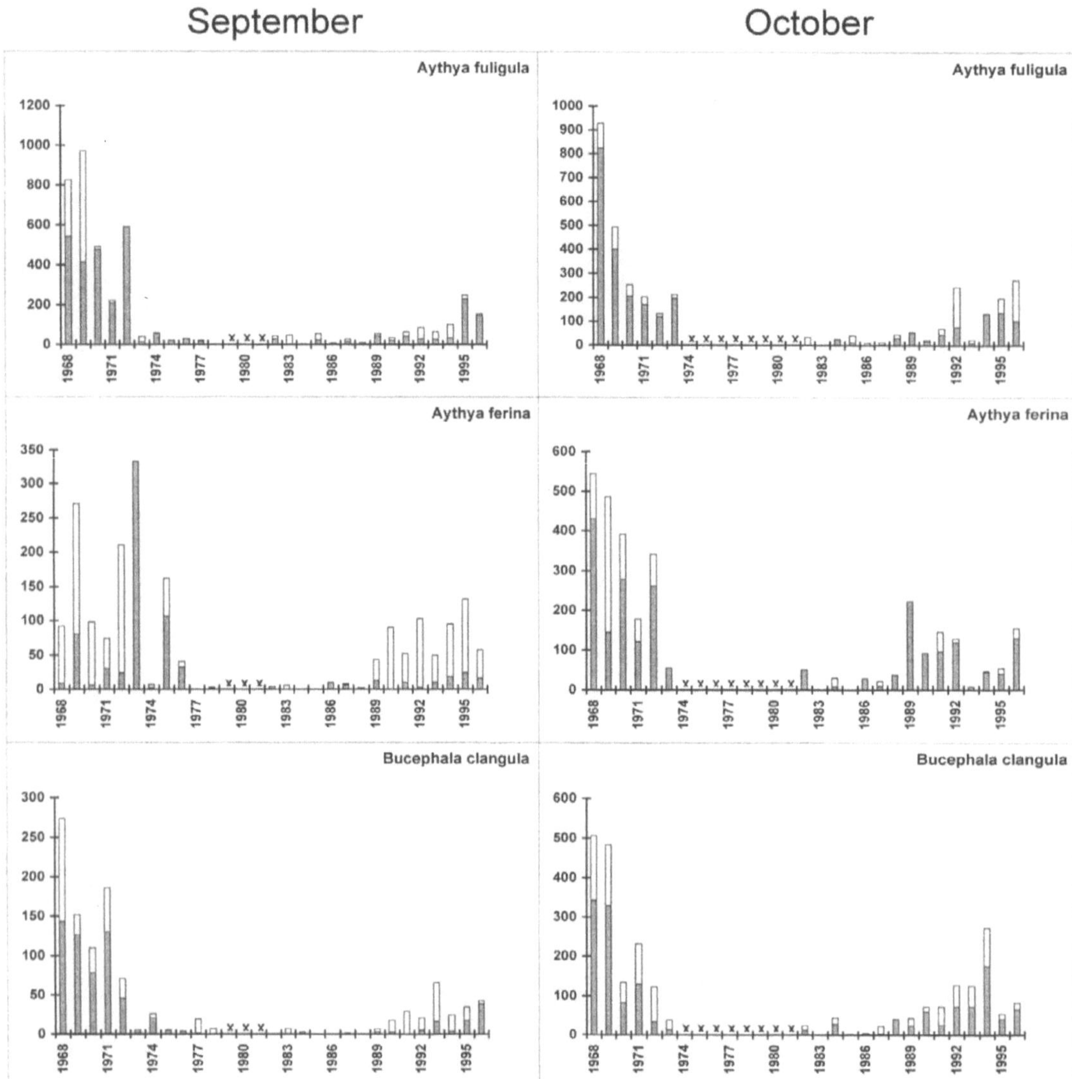

Figure 4. Number of benthivorous waterfowl counted in Western Basin (dark bar) and Eastern Basin (white bar) during September and October counts from 1968 to 1996. Benthivorous waterfowl include: the tufted duck (*Aythya fuligula*), the pochard (*Aythya ferina*) and the goldeneye (*Bucephala clangula*). Note the different scales. No counts were performed in September 1979–1981 or in October 1974–1981.

1997b). The wigeon was seen in small numbers during the first years of the counts, whereas large flocks were observed both in September and October in recent years. In the first years of the counts, the shoveler (*Anas clypeata:* Figure 3), was found in flocks of up to 290 individuals in autumn, mainly in the shallow water areas in the Fulltofta and Lillö areas (Figure 1). The abundance of shovelers decreased sharply in 1973, and remained low until 1990. After 1990, a few individuals have been observed each year.

Benthivores

When the counts began in the late 1960s, large flocks of the tufted duck (*Aythya fuligula)* were found in Lake Ringsjön (Figure 4), mainly in Western Basin, but also

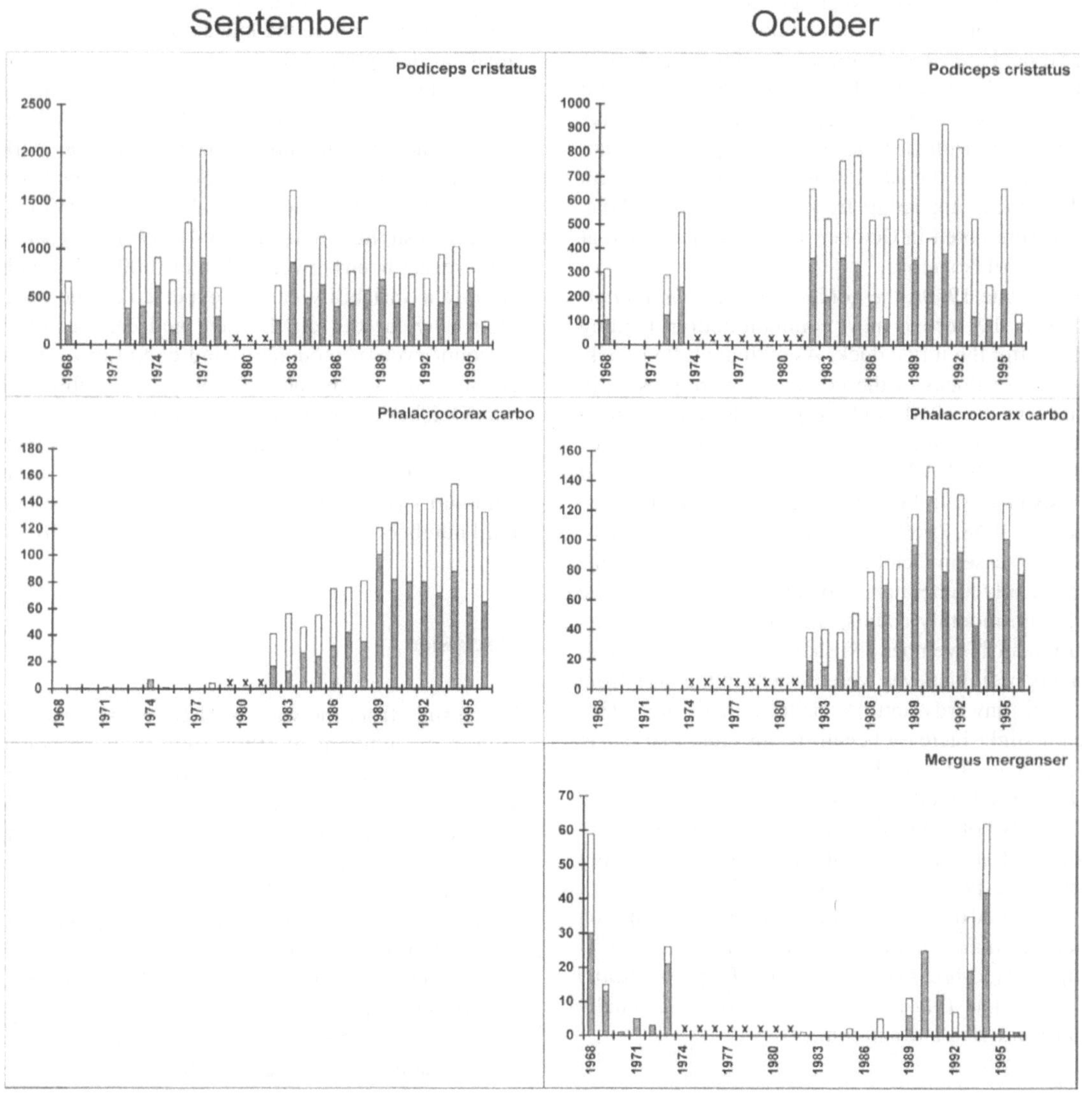

Figure 5. Number of piscivorous waterfowl counted in Western Basin (dark bar) and Eastern Basin (white bar) during September and October counts from 1968 to 1996. Piscivorous waterfowl include: the great crested grebe (*Podiceps cristatus*), the cormorant (*Phalacrocorax carbo*) and the goosander (*Mergus merganser*). Note the different scales. No counts were performed in September 1979–1981 or in October 1974–1981. Furthermore, great crested grebes were not counted in 1969–1971.

in Eastern Basin. Maximum September counts were noted in 1968 and 1969, when between 800 and 1000 tufted ducks were present. Actually, the August count in 1969 yielded even higher numbers with no less than 2200 individuals counted. In 1970, July and August counts were still 800–1000, but by 1972, late summer totals were about 600, and in 1973 only little more than 200. September totals were low from 1973 onwards. The summer flocks were generally moulting birds with high dominance of males. Large flocks of tufted ducks remained in late autumn. During all the years of the 1980s, only small numbers of tufted ducks were found in the lake. From 1988 to 1989, an increase was noted and in recent years up to 250 individuals have been counted in the autumn flocks.

The pochard (*Aythya ferina*) did not occur in late summer flocks in the late 1960s, but in September and October, flocks of a few hundreds gathered in the lake (Figure 4). As with the tufted duck, the species was not very common in the late 1970s and in the 1980s with smaller flocks appearing from 1989 onwards. Peak counts were noted in October, but flocks remained at the lake until November.

In the late 1960s, the goldeneye (*Bucephala clangula*) was the second most common bottom feeding species after the tufted duck. As with the tufted duck, late summer flocks of the goldeneye were present in the lake in the 1960s and early 1970s, Lake Ringsjön being a moulting site also for this species. August totals in 1969–1973 were 352, 158, 187, 117 and 26, respectively. In September, goldeneye flocks were present in 1968, 1969 and the first years of the 1970s, as was the case in October (Figure 4). October counts were actually higher than September counts and in the years with monthly counts, peak goldeneye numbers occurred in November (Nilsson, 1974). Only small numbers of goldeneyes were present in the lake from 1973 and onwards until the latter half of the 1980s when a slight increase began, being more marked in October than in September.

The tufted duck, the pochard and the goldeneye are the only bottom feeding diving ducks regularly occurring in Lake Ringsjön. Single individuals of other diving duck species such as the scaup (*Aythya marila*), the long-tailed duck (*Clangula hyemalis*), the common scoter (*Melanitta nigra*), the eider (*Somateria mollissima*), the red-breasted merganser (*Mergus serrator*) and the smew (*Mergus albellus*) have occasionally been recorded in late autumn.

Piscivores

The fish-eating goosander (*Mergus merganser*) is a regular visitor in Lake Ringsjön in late autumn. In the years with late autumn counts (1969–1973), peak counts amounted to 1215, 1425, 1540, 545 and 214, respectively. By September and October, only single individuals or small flocks are found in Lake Ringsjön. At the October counts, the goosander does, however, show the same general picture as the other diving ducks with flocks in 1968–1973 and after 1989, and with hardly any individuals seen in the early 1980s (Figure 5).

Two other piscivorous species are included in the data set in addition to the late autumn goosander. The great crested grebe (*Podiceps cristatus*) is the only species breeding in any numbers in Lake Ringsjön. It is also a regular visitor to the lake during all years showing no trend in numbers but merely quite marked fluctuations between different September counts (Figure 5). The highest count was about 2000 (1977) and the lowest slightly more than 200 (1996), but generally the level has been between 800 and 1000 in September. A certain variability can be expected as the species often stay far out on the lake and can be difficult to count even at moderate wind conditions. October counts in the 1980s and 1990s were noticeably higher than October counts in the first few years.

The cormorant (*Phalacrocorax carbo*) was rare in the lake during the 1970s, but has been a regular and common visitor since 1982, when an increase began until numbers stabilized around 1991 (Figure 5).

Discussion

The results from the waterfowl counts at Lake Ringsjön in the autumns of 1968–1996 show pronounced changes in numbers of most waterfowl species (Figures 2–5). When the survey started in the late 1960s, the lake was an important staging area for many species and a moulting area for the tufted duck and the goldeneye (Nilsson, 1972). In the beginning of the 1970s, the number of waterfowl decreased rapidly and within a few years many species (e.g. coot and tufted duck) almost disappeared. The abundance of waterfowl remained at a low level over a 15 year period (1973–1988). From 1989 and onwards, abundances gradually increased and for some species, the same numbers as in the late 1960s were recorded. This pattern is evident for herbivorous birds (swans and coots) (Figure 2), for omnivorous birds (dabbling ducks, but less evident for the mallard) (Figure 3) and for benthivorous birds (diving ducks) (Figure 4). Two piscivorous birds however, the great crested grebe and the cormorant, did not exhibit this pattern (Figure 5). The numbers of grebes in Lake Ringsjön varied from year to year without any specific trend, and numbers of staging cormorants in autumn increased from 1982 to 1990 and seem to have stabilized around 100–150 individuals in the 1990s. The results of the survey in Lake Ringsjön can thus be divided in three characteristic periods:

1. 1968–1972. High abundance of waterfowl followed by a rapid decline of many species.
2. 1973–1988. Low abundance of most waterfowl except fish-eating species (great crested grebe).
3. 1989–1996. Slow recovery of most species occurring in the first period of the survey.

Two potential explanations to the long-term variation in waterfowl abundances are possible: Large scale changes of the population in the region and changes in waterfowl food resources in Lake Ringsjön.

The National September Waterfowl Counts, repeated regularly since 1973, is a standardized survey of the number of waterfowl visiting lakes and coastal areas in southern Sweden (Nilsson, 1997b). It is used here as a relative measure of sizes and changes of waterfowl populations breeding in Sweden and adjacent areas. For some species (whooper swan, coot, mallard, great crested grebe and cormorant) the abundance pattern from Lake Ringsjön coincides partly with those recorded in the national counts.

The population of the whooper swan has increased during the period covered (Nilsson, 1997a). It is unlikely however, that the increase in Lake Ringsjön from near zero in the 1970s to between 30 and 120 individuals in the 1990s (Figure 2) is the result of solely a general population increase. Increased food resources in the lake and changed foraging habits may also have affected the number of staging swans. During all counts since 1989, except in October 1993, whooper swans have been present in Lake Ringsjön, foraging and resting in an area with submersed plants in Eastern Basin.

The population of coots in Sweden was severely reduced following a harsh winter in 1978/79 and did not recover until the 1990s (Nilsson, 1997b). However, the almost complete disappearance from Lake Ringsjön had already occurred by 1973 (Figure 2), and is well connected to the decline in submersed vegetation during these years (Strand, 1999).

The number of mallards in Lake Ringsjön was somewhat higher in the 1980s and 1990s compared to the first years (Figure 3), which is probably a reflexion of a general trend in the population in Sweden (Nilsson, 1997b). The September counts in Sweden also indicate an increase in the wigeon population but not to the extent found in Lake Ringsjön, indicating that local conditions are also involved.

The National Waterfowl Counts do not indicate any trends for diving ducks. In these counts there is a weak trend of decreasing numbers of great crested grebes (Nilsson, 1997b). We believe that a similar decrease in Lake Ringsjön (Figure 5) can be the result of a population decline. The cormorant showed a marked increase both in Lake Ringsjön (Figure 5) and in the National September Counts (Nilsson, 1997b). This is coincident with a general increase of the breeding population in the whole northwestern Europe (van Eerden & Gregersen, 1995).

Thus, only for mallard, great crested grebe and cormorant the common patterns between populations changes are documented by the National September Counts and the changes in number of staging birds in Lake Ringsjön (our study). We also believe that the increase in number of whooper swans in Lake Ringsjön in the 1990s may partly be the effect of a growing population. The alternative explanation for varying waterfowl abundance – changed availability of food resources – seems likely for most other waterfowl species. Information on limnological conditions in Lake Ringsjön for 1965–1974 is more scant compared to 1975 and onwards. There is, however, enough information to reconstruct events of 1965–1974. In summer, transparency decreased from about 2 m to 0.7 m and phosphorus concentration increased considerably (Bergman, 1999). Complaints about green and turbid water, and blooms of blue-green algae reached the local authorities. The earlier occurrence of rich submersed vegetation seems to have vanished around 1970 (Strand, 1999). Thus, we can conclude that Lake Ringsjön went through a rapid eutrophication process in 1969–1972. Eutrophication is a process which affects the whole ecosystem including food resources for several of the waterfowls. Hence, herbivorous birds, like swans and coots, but also some of the dabbling ducks, are directly affected when the submersed macrophytes disappear. Moreover, benthic invertebrates are important food items for diving and dabbling ducks. Nilsson (1972), reported that tufted duck and goldeneye in Lake Ringsjön utilized this resource. Several studies point out that many invertebrates are favored by stands of submersed plants (Hargeby, 1990). We can thus expect that the disappearance of submerged vegetation also resulted in an impoverishment of benthic invertebrates. The observation by Nilsson (1972) that there were a lot of empty shells of gastropods and mussels, supports this conclusion. Results from Svensson et al. (1999) also show that the benthic fauna in Lake Ringsjön was more abundant in the late 1960s compared to 1982. A likely explanation of the decrease of benthivorous birds in Lake Ringsjön after 1970 is, therefore, the reduced availability of benthic animals.

When lakes undergo eutrophication, the fish fauna changes from dominance of percid species to dominance of cyprinid species (Hartmann, 1977; Persson, 1983). This results in dense fish assemblages of often small-sized bream, roach and other cyprinids, whereas piscivorous fish are disfavoured (Persson et al., 1991). Though the water became more turbid after eutrophication this has apparently not affected the great crested grebe and the cormorant in Lake Ringsjön negatively, whereas the abundance of goosander declined (Figure 5).

Period 2 (1973–1988) was characterized by low numbers of waterfowl, except in respect to piscivorous birds (Figures 2–5). Even if no counts were performed in the lake at the middle of this period (September 1979–1981, October 1974–1981), it is likely that waterfowl abundance was at a low level. We base this assumption on the limnological conditions being fairly constant (Bergman, 1999; Cronberg, 1999). Light penetration through the water was low, and submersed macrophytes, as well as benthic macroinvertebrates, were scarce (Strand, 1999; Svensson et al., 1999). Thus, the low abundance of herbivorous and benthivorous birds in period 2 was most probably explained by reduced food resources, whereas piscivorous birds were favoured by a richness of small fish.

The increase in waterfowl abundances in period 3 (1989–1996) coincides with improved limnological conditions following fish reduction. Hence, phytoplankton biomass were reduced (Cronberg, 1999), light conditions improved and submersed macrophytes recolonized some shallow areas (own observations; Strand, 1999). Moreover, the abundance of benthic invertebrates increased considerably (Svensson et al., 1999). These changes in limnological variables resulted in improved food availabilty for most waterfowl. The increase in abundance of these birds as early as in 1989 (Figures 2 – 5) indicates that they rapidly localized and exploited the new food resources.

The conclusion that varying food resources govern the abundance of waterfowl in Lake Ringsjön is based on the fact that most foraging herbivorous, omnivorous and benthivorous birds were concentrated to shallow areas with well developed submerged vegetation. Such was the case in 1968 – 1969 also, when such birds aggregated in Lillöviken in Western Basin where both plants and invertebrates were abundant (Nilsson, 1972). During the recovery of the waterfowl in the 1990s, most foraging birds have been located at the Ormanäsviken in Western Basin and to Fulltoftaviken in Eastern Basin (Figure 1). In both areas submerged plants (mainly *Potamogeton*) have developed since 1989 (Strand, 1999). On the other hand, piscivorous birds, showing less fluctuations in population sizes, did not concentrate at these areas but foraged throughout most of the lake.

The long-term fluctuations in waterfowl abundances in Lake Ringsjön is not a unique phenomenon. Similar, but not coincident, changes are reported from nearby Lake Finjasjön after eutrophication and fish manipulation (local bird watchers; pers. comm.). In Lake Krankesjön and Lake Tåkern, Blindow et al. (1993) documented long-term waterfowl changes related to shifts in limnological conditions. Large variations in cover of submerged macrophytes can affect herbivorous birds tremendously (Hanson & Butler, 1994; Mitchell & Perrow, 1998). It is evident that submerged macrophytes are key organisms in many eutrophicated and shallow lakes. When such lakes switch between clear macrophyte dominated and turbid, phytoplankton dominated states, the whole ecosystem is influenced (Scheffer et al., 1993). Basic food resources for waterfowl may be changed with consequences for aquatic birds. In the case of Lake Ringsjön, the counter measures resulted in relatively restricted improvements in limnological conditions, although these changes were obviously great enough to affect waterfowl abundances considerably!

References

Andersson, G., 1981. Influence of fish on waterfowl in lakes. Anser 20: 21–34 (In Swedish with English summary).

Bergman, E., 1999. Changes in nutrient load and lake water chemistry in Lake Ringsjön, southern Sweden, from 1966 to 1996. Hydrobiologia 404: 9–18.

Blindow, I., G. Andersson, A. Hargeby & S. Johansson, 1993. Long-term pattern of alternative stable states in two shallow eutrophic lakes. Freshwat. Biol. 30: 159–167.

Cronberg, G., 1999. Qualitative and quantitative investigations of phytoplankton in Lake Ringsjön, Scania Sweden. Hydrobiologia 404: 27–40.

Hamrin, S. F., 1999. Planning and execution of the fish reduction in Lake Ringsjön. Hydrobiologia 404: 59–63.

Hanson, M. A. & M. G. Butler, 1994. Responses of food web manipulation in a shallow waterfowl lake. Hydrobiologia 279/280: 457–466.

Hansson, L.-A., M. Enell & E. Bergman, 1999. Lake Ringsjön: its catchment area, its history and its importance. Hydrobiologia 404: 1–7.

Hargeby, A., 1990. Macrophyte associated invertebrates and the effect of habitat permanence. Oikos 57: 338–346.

Hartmann, J., 1977. Fischereiliche Veränderungen in kulturbedingt eutrophierenden Seen. Schweiz. Z. Hydrol. 39: 243–254.

Lauridsen, T. L., E. Jeppesen & F. Østergaard-Andersen, 1993. Colonization of submerged macrophytes in shallow fish manipulated Lake Væng: impact of sediment composition and waterfowl grazing. Aquat. Bot. 46: 1–15.

Mitchell, S. F. & M. R. Perrow, 1998. Interactions between grazing birds and macrophytes. In E. Jeppesen, Ma. Søndergaard, Mo. Søndergaard & K. Christoffersen (eds), The Structuring Role of Submerged Macrophytes in Lakes. Springer–Verlag, New York.

Nilsson, L., 1972. Local distribution, food choice and food consumption of diving ducks on a south Swedish lake. Oikos 23: 82–91.

Nilsson, L., 1974. Resting and wintering waterfowl on Lake Ringsjön 1968 – 1974. Anser 13: 201–210 (In Swedish with English summary).

Nilsson, L., 1976. Monthly counts as a measure of population changes in some species of Anatidae in South Sweden. Ornis Scandinavica 7: 193–205.

Nilsson, L., 1978. Individräkning av rastande och övervintrande sjöfågel. In Statens Naturvårdsverk Biologiska Inventeringsnormer Fåglar. Stockholm (In Swedish).

Nilsson, L., 1997a. Changes in numbers and habitat utilization of wintering whooper swans (*Cygnus cygnus*) in Sweden 1964–1997. Ornis Svecica 7: 133–142.

Nilsson, L., 1997b. Internationella sjöfågelräkningarna i Sverige (In Swedish). In SOF 1997. Fågelåret 1996, Stockholm: 41–48.

Persson, L., 1983. Effects of intra- and interspecific competition on dynamics and size structure of a perch *Perca fluviatilis* and a roach *Rutilus rutilus* population. Oikos 41: 126–132.

Persson, L., S. Diehl, L. Johansson, G. Andersson & S. F. Hamrin, 1991. Shifts in fish communities along the productivity gradient of temperate lakes – patterns and the importance of size-structured populations. J. Fish. Biol. 38: 281–293.

Scheffer, M., S. H. Hosper, M.-L. Meijer, B. Moss & E. Jeppessen, 1993. Alternative equilibria in shallow lakes. Trends Ecol. Evol. 8: 275–279.

Strand, J., 1999. The development of submerged macrophytes in Lake Ringsjön after biomanipulation. Hydrobiologia 404: 113–121.

Svensson, J., E. Bergman & G. Andersson, 1999. Impact of cyprinid reduction on the benthic macroinvertebrate community and implications for increased nitrogen retention. Hydrobiologia 404: 99–112.

van Eerden, M. R. & B. J. Gregersen, 1995. Long term changes in the northwest European population of Cormorants, *Phalacrocrax carbo sinensis*. Ardea 83: 61–73.

Hydrobiologia **404**: 53–58, 1999.
L.-A. Hansson & E. Bergman (eds), Nutrient Reduction and Biomanipulation as Tools to Improve Water Quality: The Lake Ringsjön Story.

Biomanipulation in a theoretical and historical perspective

Eva Bergman[1], Lars-Anders Hansson[1] & Gunnar Andersson[2]
[1]*Department of Limnology, Insitute of Ecology, Ecology Building, University of Lund, SE-223 62 Lund, Sweden*
[2]*County Administration of Scania, SE-205 15 Malmö, Sweden*
Current address: Department of Biology, Karlstad University, S-651 88 Karlstad, Sweden

Key words: nutrient control, predator control, bottom-up:top-down, cascading trophic interactions, productivity based predator regulation theory, biomanipulation

Abstract

In limnology, nutrients have historically been regarded as main regulators of the ecosystem; that is, organisms in lakes were seen as regulated from the bottom of the food chain. Long after the seminal works of Hairston et al. (1960), Hrbáček et al. (1961) and Brooks & Dodson (1965) this 'bottom-up' view prevailed. It was not until the late 1970s and 1980s that these new theories gained recognition, considering the 'top-down' effect; that is, the impact of predation on lower trophic levels. The decision to subject Lake Ringsjön to a biomanipulation was based on the fact that the phosphorus reductions accomplished during the 1970s did not have the positive effects expected. In addition, several theories had been proposed which predicted the effects of manipulations at the top of the food chain on lower trophic levels. In this paper, we describe how these different theories influenced us at the time the studies of Lake Ringsjön were initiated, and how difficult it may be to convert scientific theories into large-scale manipulations.

Introduction

In limnology, more than in any other ecological discipline, nutrients have been regarded as the primary regulator of the structure and function of the ecosystem (Thinemann, 1921; Naumann, 1932; Rodhe, 1975). Because phosphorus is generally limiting for algal growth in lakes, this nutrient has been viewed as the main determinant for phytoplankton biomass development (Vollenweider, 1968). However, during recent years, it has been argued that top predators are of importance in regulating the abundance and size of their prey, which in turn cascade down to phytoplankton (Hairston et al., 1960; Hrbáček et al., 1961; Brooks & Dodson, 1965). This simple and appealing theory on how food chains in both terrestrial and aquatic ecosystems may function, was rapidly put into practical application. In freshwater ecosystems, the method could hypothetically lead to a lower abundance of nuisance phytoplankton, a major problem in urban, eutrophicated lakes.

In this paper, we present a short historical background behind the development of different theories for explaining lake ecosystem structure and regulation. Specifically, we will describe the concepts that influenced researchers while various attempts were made to improve water quality in Lake Ringsjön, such as 'phosphorus reduction' and 'biomanipulation'.

Nutrient regulation

There are several reasons why nutrients were viewed as the main regulating factor in lake ecosystems, and why it was the *only* view for a long period of time. One of these reasons is the idea of energy loss between adjacent trophic levels, leading to a nutrient–dependent limit for how many trophic levels a system can support. The traditional way of describing an ecosystem as a pyramid also implies that there is an upward transport of energy and matter (Figure 1). This concept was presented by Odum (1969), although he was most probably inspired by the work of Lindeman (1942), who in turn got inspiration from Elton (1927). The concept can be illustrated as a large base of primary producers, a smaller quantity of primary consumers,

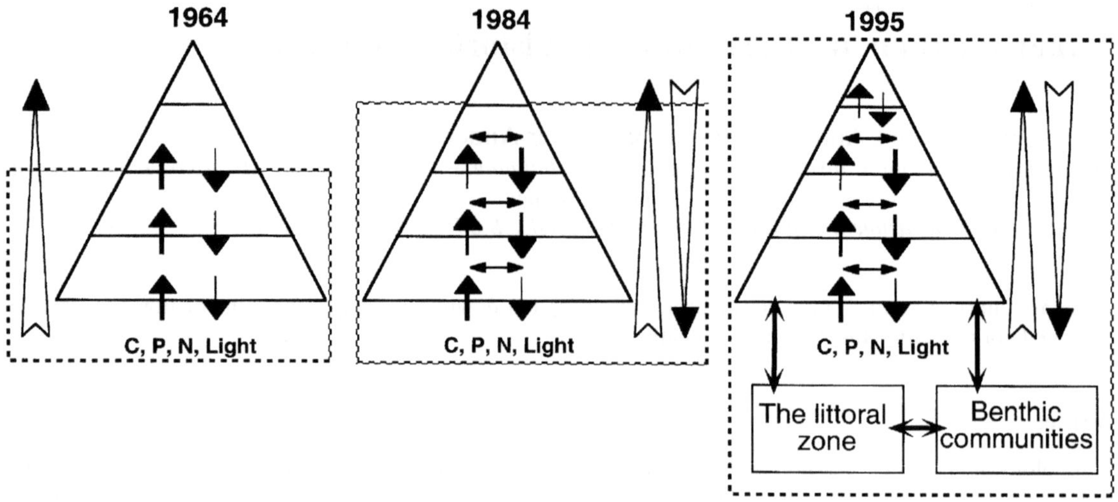

Figure 1. Illustration of the principal change in research approach used in 1964, 1984 and 1995 among limnologists working in lakes. The pyramids depict interactions in the pelagic zone; the littoral zone is not included before the beginning of the 1990s. Each pyramid is divided into four compartments representing phytoplankton, zooplankton, planktivorous fish and piscivorous fish, respectively. Large vertical arrows on the sides of the pyramids show the net direction of energy flow according to the different research approaches. Small vertical arrows show the direction and strength of interactions among compartments. Small horizontal arrows indicate the interest for competitive interactions at different trophic levels. The hatched boxes around pyramids symbolize the enlarged research fields over time, showing that in 1995 also top carnivores are included, as well as interactions with the littoral zone and the benthic community. (Partly based on Andersson, 1984).

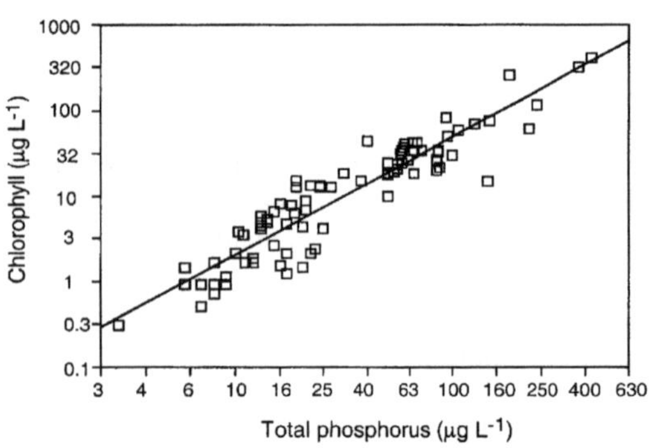

Figure 2. Chlorophyll *a* ($\mu g\ l^{-1}$) as a function of total phosphorus concentration ($\mu g\ l^{-1}$). Data from Swedish and Antarctic lakes (Hansson unpubl.).

and an even smaller quantity of secondary consumers. All are dependent and regulated by the amount of nutrients available. This pyramid concept gives the impression that the biomass of organisms in an ecosystem is a direct result of the amount of nutrients and that productivity is exclusively determined by the nutrient concentration.

Another reason for the dominance of the 'nutrient–focused view' in limnology is the indisputable relationship between the nutrient concentration and the amount of phytoplankton in a lake. In 1966, Sakamoto published regressions showing the relationship between chlorophyll content and total phosphorous. Since then, numerous studies from lakes in different parts of the world have shown a similar relationship between the amount, or loading of phosphorus and phytoplankton biomass (Figure 2) (Vollenweider, 1968; Odum, 1969; Dillon & Rigler, 1974; Nichols & Dillon, 1978; Schindler, 1978; Watson & Kalff, 1981; Hansson, 1992). Later studies also showed a relationship between phytoplankton biomass and zooplankton biomass (McCauley & Kalff, 1981; Mills & Schiavone, 1982; Canfield & Watkins, 1984; Hanson & Peters, 1984).

A third reason is related to the tradition of limnological studies, which have been focused towards physical and chemical conditions of lakes. If biotic interactions were studied, there was a strong bias towards lower trophic levels, such as phyto- and zooplankton (Persson et al., 1988; Figure 1). Interestingly, many limnologists claimed they had a holistic approach despite this obviously narrow view of aquatic ecosystems.

Moreover, there has been a lack of communication between traditional limnologists, fish biologists and microbiologists, the latter two mainly focusing on the top and bottom, respectively, of food webs. Not before the early 1980s were attempts made to integrate these different approaches (see Andersson, 1984).

Predator control – the 'food chain theory'

Although naturalists and people spending time in nature, have always had a gut feeling that organisms interact with each other and with their environment, the very beginning of a more profound and general understanding of connections within food webs was formulated as late as 1960. Then, Hairston et al. (1960) posed the question: Why is the world green? Their answer to this question, which formed the basis of classical 'food chain theory', had to do with the balance between predators, herbivores and plants. They reasoned that if green plants are abundant, herbivores graze on them, reproduce and become more abundant. Then predators, feeding on these herbivores, increase in abundance and thereby regulate the abundance of the herbivores. In this way the well-being of green plants is closely connected to the activities and abundances of predators. Hence, a never ending 'war' is going on between plants (primary producers), herbivores and predators, which constitute trophic levels 1, 2 and 3, respectively. Parallel to Hairston, Smith and Slobodkin, who developed their model for terrestrial systems, Hrbáček et al. (1961) concluded that the biomass of phytoplankton is not only a reflection of nutrient loading, but also depended on the presence of fish, which reduced zooplankton biomass, which in turn resulted in increased phytoplankton biomass. Four years later, Brooks & Dodson (1965) demonstrated how introduction of fish in a formerly fishless lake shifted the zooplankton community towards smaller zooplankton individuals and species. This phenomenon was later referred to as the 'size-efficiency hypothesis' and has proven to be a generally applicable concept. Despite these early studies, it took many years before the regulating effect of fish on lower trophic levels was fully appreciated (Figure 1). By that time, many studies had shown that introductions or removals of fish had marked effects on lower trophic levels. In a paper from 1988, Persson et al. listed 20 works from the period 1961–1986 that showed an effect of planktivorous fish on organisms in one, two or three trophic levels below them. They also pointed out that qualitatively these studies showed similar results regardless of whether the experiments were carried out in aquaria, ponds, lake enclosures or in whole lakes (Persson et al., 1988).

Integration and theory development

After several years of intellectual turbulence about whether a lake was regulated by nutrient availability or abundance of the top predator, a theory that integrated the two views of lake ecosystems was needed. An integration could simply be a theory that considered both the biotic structure of the system and the nutrient loading, such as the productivity of the system. In 1985, Carpenter and co-workers presented their concept of 'Cascading Trophic Interactions'. Parallel with this latter concept, Persson et al. (1988) developed a theory of predator regulation in relation to primary production of lakes. Both theories were based on earlier work in aquatic (Hrbáček et al., 1961; Brooks & Dodson, 1965; Shapiro et al., 1975), as well as in terrestrial ecosystems (Hairston et al., 1960; Fretwell, 1977; Oksanen et al., 1981).

The trophic cascade

The trophic cascade in lakes is based on two general principles. The first one is based on energy loss between trophic levels and the upward flow of energy through trophic levels, i.e. nutrients set the limit for primary production. As nutrients only account for some part of the variation between nutrient loading and biomass of primary producers, the remaining variation must be explained by something else. They suggested that this unexplained variation (Figure 2) was due to differences in food web composition among lakes.

The second principle implicit in the trophic cascade is based on Hrbáčeks pond findings, and Brooks & Dodsons 'size efficiency hypothesis' and suggests that a perturbation made in one trophic level will have consequences for several other trophic levels; that is, the perturbation will cascade through the food web. The cascade hypothesis states, simply expressed, that an increase in piscivore biomass causes a decrease in the planktivore biomass, an increase in herbivore biomass and a decrease in phytoplankton biomass (Figure 3). The potential productivity is established

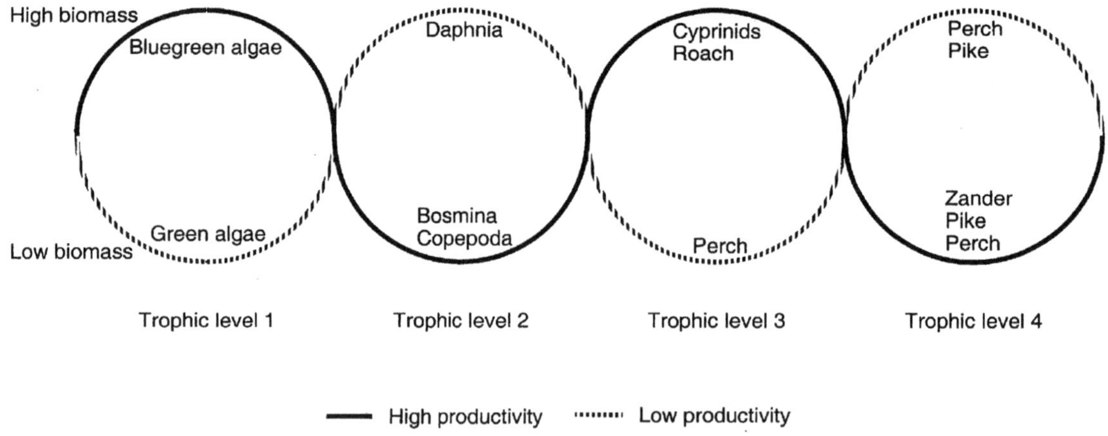

Figure 3. Illustration showing the trophic cascade in an oligotrophic (dotted line) and an eutrophic lake (solid line). The trophic cascade in a lake of high productivity has a low biomass (efficiency) of piscivores, which results in a high biomass of planktivores, low biomass of zooplankton and high biomass of algae. In a lake with low productivity the cascade begins with a high biomass of piscivores, which results in a low biomass of planktivores, high biomass of zooplankton and low biomass of algae. Note that the dominant taxon at each trophic level is given.

by abiotic factors such as nutrient supply and mixing but the actual productivity is established by biotic factors as food web structure, which in turn depend on the strength of interspecific interactions (Paine, 1980; Carpenter et al., 1985, 1987).

'Productivity based predator regulation'

Oksanen et al. (1981) developed a theory that showed how the number of trophic levels increased with increasing productivity of the system (Figure 4). Based on this, a 'productivity-based predator regulation theory' was developed for lakes by Persson et al. (1988). The theory states that lakes of low productivity can only support three trophic levels, and that the top trophic level consists of planktivorous fish. Lakes of somewhat higher productivity can support four trophic levels, with piscivores as the top trophic level. According to the theory, the importance of predation and competition alternates through the food chain such that the top trophic level is always regulated by competition, the next by predation and so on (Figure 4). An important addition made by Persson et al. (1988) relative to earlier theories was that in highly productive lakes, the visually hunting predatory fish will have problems capturing their prey due to the high algal turbidity. As a result, piscivorous fish will no longer be functionally important in regulating the abundance of planktivorous fish, and the lake functionally becomes a three-trophic level system. Hence, in very eutrophicated lakes, planktivorous fish will regulate the performance of zooplankton, because the biomass of piscivorous fish is too low to affect the abundance of planktivorous fish (Persson et al., 1988).

Bottom-up:top-down theory

Parallel to the 'cascading trophic interaction' and the 'productivity-based predation' theories, the 'bottom-up:top-down' concept evolved (McQueen et al., 1986). This concept combines the predicted influences of both predators (fish; from the top of the food chain and downwards: top-down) and resource availability (from the bottom of the food chain and upwards: bottom-up). According to this concept, nutrient availability determines the maximum attainable biomass, but the realized biomass is determined by the combined effects of bottom-up and top-down forces (see also Figure 1). In contrast to the Cascading Trophic Interaction Theory, the bottom-up:top-down concept predicts that at high nutrient concentrations, that is, in eutrophic waters, there will be no influence of fish on algae, because the system will mainly be governed by bottom-up forces. Hence, fish will be important for the development of the algal community only in low productivity (oligotrophic) systems.

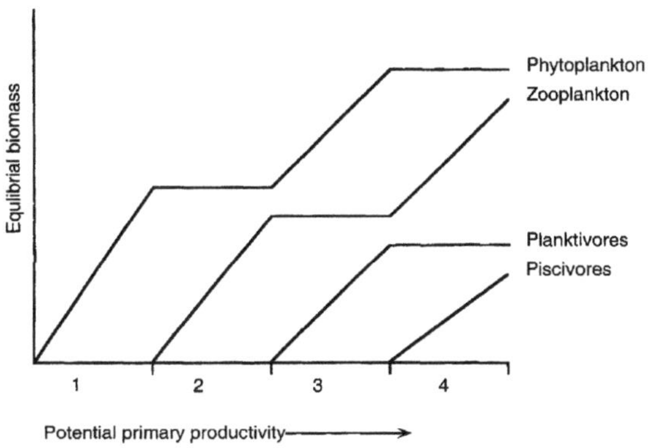

Figure 4. Graphical representation of the theory developed by Oksanen et al. (1981) showing the relation between potential primary productivity and biomass. At low productivity the energy is enough to supply one trophic level, the phytoplankton. At higher productivity the phytobiomass is high enough to maintain zooplankton in the system. As the biomass of grazers increase, phytoplankton biomass is rather constant even though production increases. At even higher values of potential primary productivity, planktivorous fish can be maintained in the system. As planktivore biomass increases, zooplankton biomass is rather constant, whereas phytoplankton biomass increases as a result of decreasing grazing from zooplankton. In systems with even higher potential primary production a fourth trophic level, piscivorous fish, can be supported by the system. The piscivores control planktivores and an increase in potential primary production should result in higher biomass of piscivores and zooplankton and not planktivores and phytoplankton. (Redrawn from Oksanen et al. (1981)).

Alternative stable states

With respect to shallow lakes, the theory developed in a somewhat different direction by also including submersed macrophytes in the models. It was also empirically found that many shallow lakes appear in a clear-water stage, where submerged macrophytes were the main primary producer, *or* in a phytoplankton dominated, turbid state *at similar nutrient status*. The theoretical implications for this phenomenon was exploited mainly by Scheffer (1990) and Scheffer et al. (1993) and has been very important for the development of modern lake management strategies.

Theory and application

The controversy between the importance of biotic interactions versus nutrient dynamics in ecosystem regulation has made clear that there is always a risk for oversimplification and misunderstanding. To some extent this is due to lack of precise definitions (see discussion in Persson et al., 1996) and different beliefs about which of the involved factors is the main regulator. When the phosphorus reductions in Lake Ringsjön were initiated in the late 1970s, the scientific discussions did not include biotic interactions, such as fish predation on zooplankton, although Hrbáček (1961) and Brooks & Dodson (1965) already had published their works. However, as early as in 1975, Shapiro and co-workers coined the term 'biomanipulation', which is an applied aspect of the food chain theory. In contrast to the 1970s, the discussions ten years later at the Department of Limnology in Lund about lake ecosystem regulation often included terms such as 'top-down effects', 'cascading effects' and 'predator-controlled systems'. One question still remained however: Was it possible to apply these theories to lake restoration?

The view of food web dynamics differs depending on if it is seen from a theoretical or management (biomanipulation) point of view. The theoretician whishes to involve all important mechanisms and let them operate in a correct way, whereas the manager simply wants to produce and maintain a clear lake. This discrepancy between theory and practice has caused much confusion and in 1998 Horppila et al. wrote: "...the original idea of biomanipulation (increased zooplankton grazing rate as a tool for controlling nuisance algae) has become a burden...". The possible background to this drastic statement is that although food chain theory is important, the concept may not embrace the whole truth. Instead, other processes, such as dynamics in littoral and benthic areas, may be as important. However, these processes are far more complex and less pedagogic to explain and are therefore only slowly implemented among lake managers.

Looking back on our experiences from Lake Ringsjön sometimes makes us feel that the theoretican and the manager live in two different worlds, and that it is very difficult to apply the theories to a whole ecosystem. Theories could possibly encourage the manager to do things that he or she otherwise would not do. When plans for large-scale measures in Lake Ringsjön were first discussed, the only theoretical basis was 'phosphorus reduction' and, later, the 'food chain' theory. As we shall see in the coming sections of this book, we had many things to learn about applying theories to real situations. However, we also feel that our work has generated several new hypotheses about factors that might be important in regulating a lake ecosystem. Thus, we are convinced that the exchange of ideas and experiences between the theoretican and the manager is crucial if progress is to be made.

Acknowledgements

We are grateful to Stephen R. Carpenter, D. J. McQueen and L. A. Greenberg for valuable and inspiring review comments. We also thank L. A. Greenberg for correcting the language.

References

Andersson, G., 1984. The role of fish in lake ecosystems – and in limnology. In B. Bosheim & M. Nicholls (eds), Nordisk Limnolog Symposium, Oslo, Interaksjoner mellom trofiske nivaer i ferskvann, 189–197.

Brooks, J. L. & S. I. Dodson, 1965. Predation, body size and the composition of plankton. Science 150: 28–35.

Canfield, D. E. Jr. & C. E. Watkins, 1984. Relationships between zooplankton abundance and chlorophyll-*a* concentration in Florida lakes. J. Freshwater Ecol 2: 335–344.

Carpenter, S. R., J. F. Kitchell & J. R. Hodgson, 1985. Cascading trophic interactions and lake productivity. BioScience 35: 634–639.

Carpenter, S. R., J. F. Kitchell, J. R. Hodgson, P. A. Cochran, J. J. Elser, M. M. Elser, D. M. Lodge, D. Kretchmer, X. He & C. N. von Ende, 1987. Regulation of Lake primary production by food web structure. Ecology 68: 1863–1878.

Dillon, P. J. & F. H. Rigler, 1974. The phosphorous-chlorophyll relationship in lakes. Limnol. Oceanogr. 19: 767–773.

Elton, C. 1927. Animal Ecology. N.Y. Macmillan Co.

Fretwell, S. D., 1977. The regulation of plant communities by food chain exploring them. Perspectives in Biology and Medicine 20: 169–185.

Hairston, N. G., F. E. Smith & L. B. Slobodkin, 1960. Community structure, population control and competition. Am. Nat. 94: 421–425.

Hanson, J. M. & R. H. Peters, 1984. Empirical prediction of crustacean zooplankton biomass and profundal macrobenthos biomass in lakes. Can. J. Fish Aquat. Sci. 41: 439–445.

Hansson, L.-A., 1992. The role of food chain composition and nutrient availability in shaping algal biomass development. Ecology 73: 241–247.

Horppila, J., H. Peltonen, T. Malinen, E. Luokkanen & T. Kairesalo, 1998. Top-down or bottom-up effects by fish – issues of concern in biomanipulation of lakes. Restoration ecology 6: 1–10.

Hrbáček, J., M. Dvorakova, V. Korinek & L. Prochazkova, 1961. Demonstrations of the effects of fish stock on the species composition and the intensity of the metabolism of the whole plankton association. Verh. Int. Ver. Limnol. 14: 192–195.

Lindeman, R. L., 1942. The trophic–dynamic aspect of ecology. Ecology 23: 399–418.

McCauley, E. & J. Kalff, 1981. Empirical relationships between phytoplankton and zooplankton biomass in lakes. Can. J. Fish Aquat. Sci. 38: 458–463.

McQueen, D. J., J. R. Post & E. L. Mills, 1986. Trophic relationships in freshwater pelagic ecosystems. Can. J. Fish Aquat. Sci. 43: 1571–1581.

Mills, E. L. & A. Schiavone, 1982. Evaluation of fish communities through assessment of zooplankton populations and measures of lake productivities. North American Journal of Fisheries Management 2: 14–27.

Naumann, E., 1932. Grundzüge der Regionalen Limnologie. Die Binnengewässer Band XI: 1–176.

Nichols, K. H. & P. J. Dillon, 1978. An evaluation of phosphorous–chlorophyll–phytoplankton relationships for lakes. Int. Rev. ges. Hydrobiol. 63: 141–154.

Odum, E. P., 1969. The strategy of ecosystem development. Science 164: 262–279.

Oksanen, L., S. D. Fretwell, J. Arruda, & P. Niemelä, 1981. Exploitation ecosystems in gradients of primary productivity. Am. Nat. 118: 240–261.

Paine, R. T., 1980. Food webs, linkage interaction strength, and community infrastructure. J. Anim. Ecol. 49: 667–685.

Persson, L., G. Andersson, S. F. Hamrin & L. Johansson, 1988. Predator regulation and primary production along the productivity gradient of temperate lake ecosystems. In S. R. Carpenter (ed.), Complex Interactions in Lake Communities. Springer-Verlag, New York: 45–65.

Persson, L., J. Bengtsson, B. A. Menge & M. E. Power, 1996. Productivity and consumer regulation – concepts, patterns and mechanisms. In G. A. Polis & K. O. Winemiller (eds), Food Webs – Integration of Patterns and Dynamics. Chapman & Hall, New York: 396–434.

Rodhe, W., 1975. The SIL founders and our fundament. Verh. Int, Ver. Limnol. 19: 16–25.

Sakamoto, M., 1966. Primary production by phytoplankton community in some Japanese lakes and its dependence on lake depth. Arch. Hydrobiol. 62: 1–28.

Scheffer, M., 1990. Multiplicity of stable states in freshwater systems. Hydrobiologia, 200/201: 475–486.

Scheffer, M., H. Hosper, M.-L. Meijer, B. Moss & E. Jeppesen, 1993. Alternative equilibria in shallow lakes. Trends in Ecology and Evolution 8: 275–279.

Schindler, D. W., 1978. Factors regulating phytoplankton production and standing crop in the world's freshwaters. Limnol. Oceanogr. 23: 478–486.

Shapiro, J., V. Lamarra & M. Lynch, 1975. Biomanipulation: an ecosystem approach to lake restoration. 85–96. In P. L. Brezonik & J. L. Fox (eds), Proceedings of the Symposium on Water Quality Management Through Biological Control. Rep. No. ENV-07-75-1.

Thienemann, A., 1921. Seetypen. Naturwissenschaften 18: 1–3.

Vollenweider, R. A., 1968. Scientific fundamentals of the eutrophication of lakes, with particular reference to nitrogen and phosphorous as factors in eutrophication. Organization for Economic, Cooperation and Development, Paris. DAS/CS1/68.27. 159.

Watson, S. & J. Kalff, 1981. Relationships between nannoplankton and lake trophic status. Can. J. Fish Aquat. Sci. 38: 960–967.

Hydrobiologia **404:** 59–63, 1999.
L.-A. Hansson & E. Bergman (eds), Nutrient Reduction and Biomanipulation as Tools to Improve Water Quality: The Lake Ringsjön Story.

Planning and execution of the fish reduction in Lake Ringsjön

Stellan F. Hamrin
Institute of Freshwater Research, SE-178 93 Drottningholm, Sweden

Key words: methods of fish reduction, biomanipulation, boats, trawls, costs

Abstract

Lake Ringsjön, a naturally productive lake in southern Sweden, was subjected to high external phosphorus loading during the 20th century and became highly productive. In spite of effective waste water treatment which reduced external phosphorus loading, the improvements in lake water quality were moderate and proceeded at low rate. Consequently cyprinid reduction, which was expected to affect water clarity via several intertrophic mechanisms, was implemented as a measure for restoring the lake. The fish removal was performed using a board trawler equipped with a herring trawl modified for bottom trawling. The bottom rope of the trawl was 11 m and the total length of the trawl was 20 m, the cod end mesh size was 11 mm (knot to knot). In the beginning the catch amounted to 1 ton of fish in less than 2 hours, logistic problems bringing the catch to shore and replacing all piscivorous fish into the lake reduced the initial daily catch to 2–3 tons. A grand total of 240 tons of fish were removed by trawling Lake Ringsjön over a 3 year period. The catch was either fermented with myric acid or deep-frozen before being used as mink and pelican food respectively. Some of the catch was also used for gas production. The cost for boats and equipment was approximately 60 000 US$. The three persons needed for fishing and sorting as well as the costs of taking care of the catch on land are not included in this sum.

Introduction

The long term gradual eutrophication of Lake Ringsjön accelerated during mid-20th century due to urbanisation and intensified agricultural activities in the drainage area. Awareness of the problem led to improved waste water treatment practices. In spite of these measures, transparency of Lake Ringsjön was still very low, the lake was dominated by blue-green algae and had a fish community dominated by stunted, slow-growing populations of roach and bream. This failure led to increased action against phosphorus-leakage from intensified agriculture activities, which during the 1980s resulted in a considerable reduction of the external phosphorus load. In spite of a lower phosphorus concentration in the water and a reduction in algal biomass, the lake showed little improvement with respect to water quality. It was thus decided that further measures must be taken.

In the 1960s, Lake Trummen (1 km^2), located 300 km north of Lake Ringsjön, was restored by sediment and vegetation removal (Björk, 1972, 1994). Although the sediment dredging in Lake Trummen was relatively successful, Lake Ringsjön was regarded as too large to be suitable for this method. However, experiences from lake Trummen proved instrumental in identifying another type of lake restoration measure, namely that immigration of cyprinids counteracted the restoration efforts. It was demonstrated that high densities of cyprinids increased phosphorus concentrations, favoured blue-green algae and subsequently reduced water transparency. Later studies of fish production and ecosystem effects of fish reductions suggested that top-down effects may be the key to successful lake restoration (Andersson, 1984; Persson, 1985; Persson et al., 1993).

Based on models by Hairston et al. (1960), Fretwell (1977), Oksanen et al. (1981) and Carpenter et al. (1985), it was hypothesized that four trophic level lake ecosystems subjected to increased external phosphorus-input developed functionally into 3-level systems, where fish piscivores lost their ability to structure the ecosystem (Persson et al., 1988). In documented cases like Lakes Ringsjön and Finjasjön, the external phosphorus input had exceeded 1 g phosphorus m^{-2} yr^{-1}. When large amounts of phosphorus were deposited in the sediments, selective predation and phosphorus-metabolism induced by cyprinid fish could strongly counteract effects of restoration through reduction in external phosphorus inputs (Sverdrup et al., 1991).

The status of Lake Ringsjön became a politically important question during the 1980s since the lake did not improve despite heavy investments (>15 million US$) in sewage water treatment plants and extensive restrictions for agricultural activities. Consequently, researchers at the Institute of Limnology at Lund University became involved in Lake Ringsjön restoration. They proposed reducing the biomass of cyprinids in the Lake Ringsjön by trawling. The main hypothesis was that removal of cyprinids would decrease predation on large zooplankton and reduce internal phosphorus circulation, thereby reducing problems with blue-green algal blooms, increasing water transparency and altering the fish community so that percids would dominate. This proposal led to strong opposition from a number of different groups claiming that the hypothesis was flawed, a sufficient amount of fish could never be removed or an increase in cyprinid juvenile survival would rapidly counteract the proposed effects of the removal.

The cyprinid removal project was supported by the National Board of Fisheries (NBF) and by local fishermen. Eventually the county board approved the project, and financial support was obtained from the National Environmental Protection Board and by local authorities from Eslöv, Höör and Hörby. The project commenced in the spring of 1989, with the launch of the trawler R/V Rödingen, which was donated to the project by the National Board of Fisheries.

Study site

Lake Ringsjön is situated in southernmost Sweden (55° 52 N; 13° 32 O) and consists of three basins: Sätofta Basin (4.2 km^2, mean depth 3.0 m, maximum depth 17.0 m), Eastern Basin (20.5 km^2, mean depth 6.1 m, maximum depth 16.4 m) and Western Basin (14.8 km^2, mean depth 3.1 m, maximum depth 5.4 m) (Hansson et al., 1999). As many other lakes in urban and agricultural areas, Lake Ringsjön became eutrophicated during the 1960s and 1970s. In the beginning of the 1980s, a *nutrient reduction programme* was initiated which reduced the external input of nutrients to the lake from phosphorus loadings exceeding 30 tons per year to around 10 tons per year. About ten years later (1989–1990), a *cyprinid fish reduction programme* (biomanipulation) was carried out in Sätofta Basin removing about 100 tons of fish (about 60% of the standing stock). During the winter of 1988, there was a fish-kill in Eastern Basin, eliminating about 500 tons of fish (about 80%). The biomanipulation in Western Basin, which was performed in 1992, removed about 120 tons of fish (about 50%).

Lake Ringsjön is naturally highly productive and has been affected by agricultural activities for quite some time (Trybom, 1893). It has been an important lake for fishing activities for at least 2 000 years. Pike (*Esox lucius*), perch (*Perca fluviatilis*), eel (*Anguilla anguilla*), bream (*Abramis brama*), roach (*Rutilus rutilus*) and planktonic white-fish (*Coregonus oxyrhynchus* sensu Svärdsson) and pike-perch (*Stizostedion lucioperca*) have been exploited from this lake. Pike-perch was first introduced in the lake during the 1930s.

Thanks to detailed descriptions by Filip Trybom (1893), our knowledge about Lake Ringsjön at the end of the last century is good. Even if blooms of phytoplankton occured, all signs indicate that Lake Ringsjön was a lake structured by piscivorous fish like perch and pike. Water transparency was fairly high, approximately 2 m, and large areas in Western Basin and Sätofta Basin were covered by submerged vegetation, forming suitable habitats for numerous species of crustaceans and molluscs, many of which later almost became extinct.

The fishing in 1883 was based on eel, bream, pike and white-fish. The cyprinid fish community was characterized by extremely large-sized bream (up to 8 kg), famous for its quality all over Scania. Using 45 boats with sails and oars, 14 seine nets, 637 gill nets, 54 fyke nets and 11 000 long-lines ('eel-lines'), 20 fishermen and 40 'lads' caught 4 200 lispund (35 000 kg = 7.5 kg/ha) anually.

At the end of the 19th century, the zone of submerged vegetation expanded as a result of water-lowering. "Since the water-lowering an enormous bed of vegetation has developed in the lake, since Lake Ringsjön is very shallow in relation to it's size. During summer you can see enormous masses of water vegetation floating on the surface, forming real virgin forests of several acres in width." (Fisherman H. Nilsson 1914; translated).

Fish removal methods

Freshwater trawling by small boats was first used in Lake Ivösjön (300 km north of lake Ringsjön) in the 1970s to catch juvenile smelt in deep water. Later on, trawling was used in Lake Sövdeborgssjön (50 km south of lake Ringsjön) for catch-recapture experiments (Persson et al. 1993). These experiences formed the basis of the trawling in Lake Ringsjön (Hamrin, 1993).

The trawl used in Lake Ringsjön was a modified pelagic herring trawl, with a bottom rope of 11 m, a trawl mouth diameter of 7 m and two 5 m long wings. The length of the trawl was 20 m and the mesh decreased from 36 mm (knot to knot) at the mouth to 11 mm at the cod end, a 5 mm cod end was used on several occasions. All trawls were made by 'Västkustens trålbinderi AB' in Glommen or Simrishamn, Sweden.

The trawl was dragged along the bottom and was thus equipped with extra weights (about 10–15 kg) along the botton rope and with 3 small floating devices (d. = 15 cm) along the upper one. During method development, the trawl was dragged at a speed of 2 knots by two 4.5–5 m plastic boats powered by 15–20 hp outboard engines with low speed propellers. The length of the lines were adjusted to the depth (2–15 m) and varied between 50 and 150 m.

Test trawlings by small boats in Lake Ringsjön in 1987 showed that this method resulted in catches of up to 600 kg of cyprinids per hour of trawling. Thus, the problem was not how to catch the fish but how to handle the catch in a cost-efficient way. Subsequent trawlings were conducted using the 12 m wooden trawler 'R/W Rödingen', an 11 ton boat equipped with a 125 hp engine.

Test runs showed that maximum efficiency in catching cyprinids was achieved at a speed of 2 m s^{-1}. Increased speed did not increase the catch and in fact the mean catch value at speeds > 2 m s^{-1} were lower. Additional trials showed that the larger mesh-sized (11 mm) cod end was three times more effective at capturing cyprinids than the cod end with 5 mm mesh size at speeds of 2 m s^{-1} (t-test, $p < 0.05$, $n = 10$).

Handling of the fish

The main problem with fish reduction was landing the cyprinids and returning all other fish species to the lake. The procedure was ultimately carried out as follows: The cod end was lifted hydraulically and emptied into two large cylindrical nets placed in an open plastic 5 m boat attached to the side of the trawler. The trawl was immediately set again, while piscivorous pike, perch and pike-perch were returned to the lake. When the nets were filled with fish (about 400 kg each) the small boat returned directly to the harbour, where the nets were lifted hydraulically and emptied into a 1 m^3 container.

The most difficult problem to solve was disposing of the catch in an economically and ethically acceptable way. Most of the fish were used for animal food, although it was not possible to use the fish for fish meal due to the long distance to the nearest factory.

The original idea was to sell the frozen fish to fox farms, but the collapse of the nearby industry at this time made this impossible. There was no interest from that industry, nor were mink farms interested because of the high concentrations of thiaminase in the fish. The only solution was fermenting the catch, and by doing this eliminating the thiaminase. This was accomplished by putting the fish into 1 m^3 containers together with an appropriate amount of formic acid and stirring the contents for approximately 12 hours. The product was a relatively thick, brown soup with a slightly acidic taste that was storable for at least 1 year. Scales and bones formed small unsoluable remains in the container.

Preliminary tests showed that the cyprinid soup might be an acceptable product for pipe-line fed pigs. Due to former misuse of (fatty) herring for this purpose, fish are not allowed as food for pigs in Sweden except in very small quantities. The industry, however, showed an interest in testing this product, but the 1 year of testing required, made this option unavailable. Instead, the cyprinid soup was donated to mink farms, as food for minks. Frozen roach and small bream were purchased (about 1 US$ per 10 kg) by central European zoological gardens, where they were used as food for pelicans!

Fishing

During the autumn of 1989 and the spring of 1990, approximately 100 tons of roach and bream were removed from the lake. Fishermen using large eel fyke nets caught an additional few tons. All piscivorous fish were returned to the lake without any signs of mortality. A few whitefish were caught and all of these died.

Trawling in Sätofta Basin was most difficult. On several occasions low water hindered the boat from entering Sätofta Basin from Eastern Basin where the boat was docked. The circular shape of Sätofta Basin restricted the duration of trawl drags and increased the non-trawling time. During some periods shoals of cyprinids concentrated in the small, deep part of the basin, where they were partially out of reach of the trawl. The CPUE (catch per unit effort) in the test trawlings was reduced by 80% after catching 100 tons, even if fish immigration from Eastern Basin later counteracted this effect to some extent.

During the spring of 1988 a fish-kill occurred in Lake Ringsjön as well as in many other lakes in Scania. Based on the number of dead fish found along the shore the fish-kill was considered to be of importance. The considerable impact of the fish kill was probably due to a number of co-occurring factors: The cyprinid biomass in the lake was large, resulting in food shortage and poor condition of individual fish. In the test trawlings of 1987 it was noticed that most of the bream and many of the large roach were thin and had infections characterized by several reddish 5–15 mm large round wounds without scales. The winter 1987/88 was unusually mild, increasing fish metabolism. The spring was very late, but in May the temperature increased rapidly. It is therefore believed that the fish kill was due to food shortage, in combination with high temperatures and infections, and affected most part of the cyprinid population.

The conclusion that there had been a fish kill was supported by test trawlings in the large Eastern Basin in 1990, when the CPUE was reduced by far more than 50% compared to 1987. Based on the relation between the catch and the test fishing results in the Sätofta Basin the reduced CPUE in Eastern Basin equals approximately 500 tons of cyprinids. As a consequence, fish removal was not neccessary in Eastern Basin. In test fishings in 1990, no signs of infection were observed, presumably because the infected individuals had died.

The fish removal in Western Basin was straightforward once the trawler had been taken through the very narrow and shallow sound and under the bridge between Eastern and Western Basins. Approximately 100 tons of roach and bream were also caught in Western Basin. In this basin however, the fishing with fyke nets was more efficient and accounted for about 20 tons of bream and roach, the total catch therefore amounted to 120 tons. Also in this basin all piscivirous fish were returned to the lake.

Concluding remarks

I would like to draw some conclusions from the cyprinid reduction in Lake Ringsjön: To reduce the likelihood that the ecosystem will return to its eutrophic state I believe it is important, if possible, to complete the fish reduction within 1 year. If not, recruitment of younger cyprinids will increase, thereby decreasing water transparency.

With respect to the fish community, it must be stressed that fish are extremely mobile creatures and individual fish will respond quickly to changing conditions. They are very likely to avoid fishing by moving to undisturbed areas like streams and vegetation. They are also, when given the possibility, very likely to move to areas with lower fish abundance. This will, however, only occur when the water transparency and the predation risk between the different habitats is equal. It is therefore important to monitor the fish abundance in different parts of the lake during fish removal.

Efficient fishing and handling of the catch is vital to keep down costs. When the cyprinid reduction was continued in the nearby Lake Finjasjön (Annadotter et al., 1998), the fishing was made more efficient, using two specially designed trawlers. The trawl was then dragged between the two boats instead of board trawling. The existence of two trawlers also has the advantage of a lesser risk of having to stop the fishing all together since one trawler can temporarily be replaced by a small outboard boat. Two trawlers also have a much larger capacity to store the catch before returning to the quay. The foredeck of each of these trawlers was equipped with a 2×2 m sorting board, on which all piscivorous fish were easily collected. The cyprinids were then released into 500 l containers placed under board through an adjustable opening in the center. After the catch, the containers were directly unloaded from the trawlers to lorries, which brought the catch to its final destination.

Using paired trawlers reduced costs associated with the boats to approximately 70 000 US$ per year, including depreciation, rents, energy and repairs. At least 3 persons are needed on the boats when fishing, and depending somewhat on the fish biomass, approximately 10 km^2 of lake can be trawled within one year (9 months of fishing). An efficient procedure also demands reliable means of transportation for the landed fish.

Acknowledgement

This project was made possible through the scientific knowledge developed by Jan Stensson, Gunnar Andersson and Lennart Persson and by the discussions with my other colleagues Eva Bergman, Lars Collvin, Lars-Anders Hansson, Lars Johansson and Olof Lessmark. The trawling was conducted by the skippers, Sven Hartwall, Åke Möller and Ebbe Nilsson, and by the 'lads', Peter Eklöv, Marie Eriksson, Henrik Hamrin, Tomas Honeth, Jan Johansson, Jesper Leijerstam and Henric Linge. The project would never have started without the initial grant from the Swedish National Board of Fisheries through Åke Petersson. Grants were later also received from the National Environmental Protection Board, from the county board of Malmöhus län and from the local authorities of Eslöv, Höör and Hörby. The project was also made possible by the continuous engagement and support from the fisherman John-Ove Bergström and his co-workers and would never have started without his participation, which is greatly appreciated. The, sometimes arduous, work was made far more pleasant by the fortifying fish dishes prepared by Bertil.

References

Andersson, G., 1984. The role of fish in lake ecosystems – and in Limnology. In S. Bosheim & M. Nicholls (eds), Interactions Between Trophic Levels in Fresh Water. Norsk limnologforening, Oslo: 189–197.

Annadotter, H., G. Cronberg, R. Ågren, L. Jönsson, B. Lundstedt, P.-Å. Nilsson & S. Ströbäck, 1998. Multiple techniques for the restoration of a hypereutrophic lake. In D. Harper, W. Brieley, G. Phillips, & A. Ferguson (eds), The Ecological Basis for Lake and Reservoir Management. Wiley, Chichester.

Björk, S., 1972. Ecosystem studies in connection with restoration of lakes. Verh. Int. Ver. Limnol. 18: 379–387.

Björk, S., 1994. In M. Eiseltová (ed.), Restoration of Lake Ecosystems – a Holistic Approach. IWRB Gloucester: 1–173.

Carpenter, S. R., J. F. Kitchell & J. R. Hodgson, 1985. Cascading trophic interactions and lake productivity. Bioscience 35: 634–639.

Fretwell, S. D., 1977. The regulation of plankton communities by the food chains exploiting them. Perspect. Biol. Med. 20: 169–185.

Hairston, N. G., F. E. Smith & L. B. Slobodkin, 1960. Community structure, population control and competition. Am. Nat. 94: 421–425.

Hamrin, S. F., 1993. Lake restoration by cyprinid control in Sätofta Bay (Lake Ringsjön). Verh. Int. Ver. Limnol. 25: 487–493.

Hansson, L.-A., M. Enell & E. Bergman, 1999. Lake Ringsjön: its catchment area, its history and its importance. Hydrobiologia 404: 1–7

Nilsson, H., 1914. Något om fisket i Ringsjön (Some notes about the fishery in Lake Ringsjön). (In Swedish) Svensk Fiskeritidskrift 23: 80–82.

Oksanen, L., S. D. Fretwell, J. Arruda & P. Niemala, 1981. Exploitation ecosystems in gradients of primary productivity. Am. Nat. 118: 240–261.

Persson, L., 1985. Asymmetrical competition – are larger animals competitively superior? Am. Nat. 26: 261–266.

Persson, L., G. Andersson, S. F. Hamrin & L. Johansson, 1988. Predator regulation and primary production along the productivity gradient of temperate lake ecosystems. In S. R. Carpenter (ed.), Complex Interactions in Lake Communities. Springer Verlag, N.Y.: 45–65 p.

Persson, L., L. Johansson, G. Andersson, S. Diehl & S. F. Hamrin, 1993. Density dependent interactions in lake ecosystems: whole lake perturbation experiments. Oikos 66: 193–208.

Sverdrup, H., P. Warfvinge & S. F. Hamrin, 1991. A simple model for the eutrophication of Lake Ringsjön. Vatten 47: 197–203.

Trybom, F., 1893. Ringsjön i Malmöhus län dess naturförhållande och fiske. (In Swedish) Medd. Kongl. Lantbruksstyr. 4: 1–42.

Hydrobiologia **404:** 65–75, 1999.
L.-A. Hansson & E. Bergman (eds), Nutrient Reduction and Biomanipulation as Tools to Improve Water Quality: The Lake Ringsjön Story.

The effects of cyprinid reduction on the fish community

Eva Bergman[1], Stellan F. Hamrin[2] & Pia Romare[1]
[1]*Department of Limnology, Insitute of Ecology, Ecology Building, University of Lund, SE-223 62 Lund, Sweden*
[2]*Institute of Freshwater Research, SE-178 93 Drottningholm, Sweden*
Present address: Department of Biology, Karlstad University, S-651 888 Karlstad, Sweden

Key words: fish community, fish reduction, biomanipulation, piscivores, planktivores

Abstract

Lake Ringsjön in southern Sweden was restored by biomanipulation by means of a reduction in zooplanktivorous cyprinids. The fish community in the three different Basins of Lake Ringsjön were followed between 1987 and 1994. In 1990, c. 100 tons (60% of the community) were removed from Sätofta Basin, in 1988, c. 500 tons (80%) were removed from Eastern Basin and in 1992, c. 120 tons (50%) of cyprinids were removed from Western Basin. In all basins, the fish biomass was much lower after fish reduction than before, but only in Eastern Basin has the proportion of piscivores increased substantially (from 10% to 25–40%). Due to very high abundance of young-of-the-year following the fish reductions, the fish abundance has increased in all basins. The abundance of the young-of-the-year was negatively correlated to the abundance of piscivore perch at the 3 m depth. Although the young-of-the-year abundance was very high after the fish reduction, there are no clear evidences that this affects recruitment to the adult populations. The results are discussed in relation to critical fish biomasses and critical proportions of planktivory and piscivory to ensure a long lasting effect of fish reduction.

Introduction

Fish communities in high productivity European lakes are usually dominated by cyprinids, mainly roach and bream (Hartman & Nümann, 1977; Leopold et al., 1986; Persson et al., 1991), and populations of these species are in turn dominated by small, young individuals (Lessmark, 1983; Persson et al., 1988; Hamrin et al., 1991). This dominance of cyprinids, many of which are planktivores, results in high predation pressure on the zooplankton community, and consequently a low grazing pressure on phytoplankton. The high biomass of cyprinids may also play an important role in conserving lakes at a high productivity stage, even after nutrient inputs to the lakes are reduced (Andersson et al., 1978; Henrikson et al., 1980; Brabrand et al., 1986). This is due to a dense cyprinid fish releasing large amounts of phosphorus via excretion and benthic feeding, thereby maintaining high algal biomass and productivity in these lakes (Andersson et al., 1978; Brabrand et al., 1990; Horppila & Kairesalo, 1990; Persson, 1997).

During the early 1980s, scientists started an intense discussion on top-down effects in ecosystems and lately various theories about top-down control and cascading effects have developed (Oksanen et al., 1981; Carpenter et al., 1985; Persson et al., 1988). These theories renewed our ideas about lakes as ecosystems and many forces wanted to put them to the test in large scale situations. Some large scale experiments have been performed, generally assuming an effect from the upper trophic levels on the lower trophic levels, e.g. in Lake Mendota, Lake Vesijärvi and some Dutch and Danish lakes (Horppila & Kairesalo, 1990; van Donk et al., 1990a,b; Søndergaard et al., 1990; Riemann et al., 1990; Jeppesen et al., 1990a,b; Kitchell, 1992). These whole lake experiments have revealed some of the gaps between theory and reality, and consequently the scientific discussion has come to include not only the pelagic system but also the role of vegetation and interactions between littoral and pelagic systems (Schriver et al., 1995). Another discussion running at the time this study began dealt with the importance of having low external phosphorous input at the start of the biomanipulation in order to achieve a positive, long lasting result (Jeppesen et al., 1990a,b; Hamrin, 1993).

This chapter describes the changes in the fish community of the three basins of Lake Ringsjön following the cyprinid reduction. We will focus on the species composition and size of the fish community, which were followed during a period of eight years, beginning 1.5 years before the first cyprinid reduction.

Study site

Lake Ringsjön is situated in southernmost Sweden (55° 52′ N; 13° 32′ O) and consists of three basins: Sätofta Basin (4.2 km^2, mean depth 3.0 m, maximum depth 17.0 m), Eastern Basin (20.5 km^2, mean depth 6.1 m, maximum depth 16.4 m) and Western Basin (14.8 km^2, mean depth 3.1 m, maximum depth 5.4 m) (Hansson et al., 1999). As is the case with many other lakes in urban and agricultural areas, Lake Ringsjön became eutrophicated during the 1960s and 1970s. In the beginning of the 1980s, a *nutrient reduction programme* was initiated which reduced the external input of nutrients to the lake from phosphorus loadings exceeding 30 tons per year to around 10 tons per year. About ten years later (1989–1990), a *cyprinid fish reduction programme* (biomanipulation) was carried out in Sätofta Basin removing about 100 tons of fish (about 60% of the standing stock). During the winter of 1988, there was a fish kill in Eastern Basin, eliminating about 500 tons of fish (about 80%) (Hamrin, 1999). The biomanipulation in Western Basin, which was performed in 1992, removed about 120 tons of fish (about 50%).

Materials and methods

In 1987, 1990, 1991, 1992, 1993 and 1994 test trawlings were performed in Sätofta Basin, Eastern Basin and Western Basin in August, except for Eastern basin in 1987 when the trawling was performed in September. We trawled at five depths in Sätofta Basin (2, 3, 4, 5 and 7 m) and in Eastern Basin (3, 5, 7, 10 and 13 m) and at four depths in Western Basin (2, 3, 4 and 5 m). The fishing was performed by pair benthic trawling (two boats are pulling a trawl) and the trawl we used was a reduced herring trawl (total length = 30 m, arm length = 5 m, and 264 meshes along the opening). The smallest mesh size in the cod-end was 5 mm in 1987 and 13 mm in the other years. A test of the two trawls was performed in Sätofta Basin in 1990 over a large 2 m plateau. The results showed that the 13 mm trawl was 3.0 times as efficient as the 5 mm trawl concerning roach and bream (which were the only species catched), consequently the catches from 1987 were multiplied by a factor of three in all our calculations. The trawling speed was ≥ 1 m s^{-1} ($\sim$2 knots). Earlier tests showed that catches of roach and bream are independent of speed when the speed is at least 1.5 knots. Each depth was trawled twice for 10 min on each sampling occasion. If the area at a certain depth was too small for 10 min trawling, we trawled as long as possible and corrected for time in the calculations afterwards. The fish in the catch were identified by species, counted, measured (length) and weighed. For very large catches we took a subsample from the total catch of dominant species. All fishing was performed between 9.00 and 17.00 hrs.

Total catch is presented as catch per unit effort (CPUE), both as biomass (kg 20 min^{-1}) and as abundance (No. 20 min^{-1}) and is based on the catches from all depths. The mean values were corrected for the relative depth distribution, which was done because we trawled similar time intervals at all depths resulting in that the total effort was larger at deeper than at shallower depths. To obtain a more realistic total catch for each species, we corrected for the percentage of total lake area represented by each depth. All statistical analyses are based on these data.

To determine whether or not the composition of the fish community altered after the fish reduction, we performed a discriminant analysis (DA) of all species in the catches from 1987 to 1994 before and after the fish reduction. Since the fish reduction occurred at different years in the different basins, the number of years before and after the fish reduction varied among basins (1990, 1988, 1992 in Sätofta Basin, Eastern Basin and Western Basin, respectively). We also analysed changes in biomass and abundance using linear regressions of the four most common species, roach, bream, perch and zander, in each basin separately. These species were also divided into juvenile and adult categories to examine recruitment, as well as to estimate the proportion of planktivores vs. piscivores in the fish community. Fish smaller than 70 mm (roach, bream and perch), and 100 mm for zander were grouped as juveniles. Regressions were performed to analyse changes over time for individual species or group of species. All data were ln-transformed before analyses except the proportion of piscivores (%) which were arcsin-square-root transformed.

Results

The biomass of perch, roach, bream, zander and other species (whitefish, burbot, eel and pike) were analyzed using discriminant analysis (DA) from 1987 to 1994. DA successfully discriminated 'before' and 'after' the fish reduction in the three basins in 94% of the cases ($F_{5,12} = 4.925$, $P = 0.011$). The biomasses of bream and roach, which both decreased after the fish reduction, were best at separating the two periods (i.e. had the highest loadings). Separate DA of the abundance of fish was less successful than similar calculations based on biomass at discriminating the period 'before' and 'after' the fish reduction ($F_{5,12} = 2.629$, $P = 0.079$). However, DA separated the periods in 94% of the cases, and none of the species had loadings over 0.5. Where the fish divided into juveniles (<70 mm) and adults (>70 mm), DA's were even worse at discriminating the two periods, only 67% were correct for fish <70 mm and 83% for fish >70 mm (<70 mm, $F_{4,13} = 1.228$, $P = 0.346$ and >70 mm, $F_{5,12} = 2.338$, $P = 0.106$). Roach and perch were best at discriminating the two periods for fish <70 mm and roach and bream for the fish >70 mm. The juvenile perch and roach increased and the adult bream and roach decreased following fish reduction.

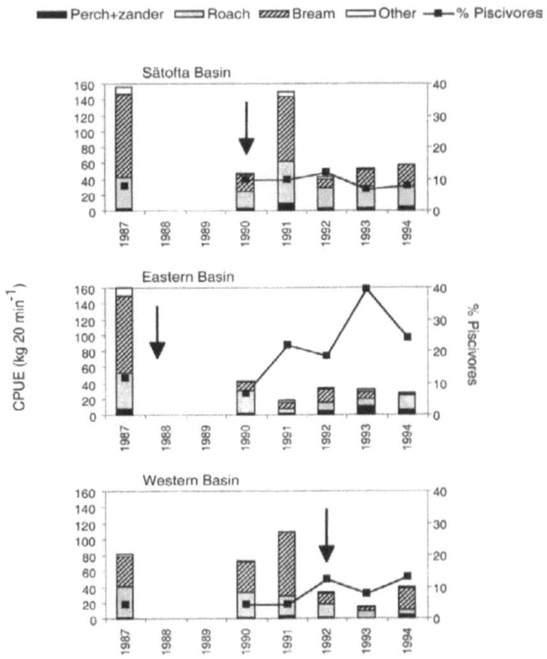

Figure 1. Fish biomass (CPUE, kg 20 min^{-1} trawling) in different years in Sätofta Basin, Eastern Basin and Western Basin. Arrows indicate year of fish reduction.

Community composition

During the whole period 1987–1994 the biomass of roach (18–62%) (minimum–maximum values) and bream (5–68%) dominated the fish catches in all basins (Figure 1). The amount of piscivores increased in Eastern Basin (6–39%) and Western (4–13%) Basin, but not in Sätofta Basin (6–12%) (Regression, arcsin-square-root-transformed data, $F_{1,4} = 0.354$, $P = 0.584$ in Sätofta Basin, $F_{1,4} = 5.109$, $P = 0.087$ in Eastern Basin, $F_{1,4} = 7.674$, $P = 0.05$ in Western Basin). In Sätofta Basin, the catches were about three times as large before fish reduction, as after the reduction. However, there was no statistically significant decrease over the whole period 1987–1994 (Regression, Ln-transformed values, $F = 2.3$, $P = 0.203$) because of the high fish biomass in 1991. This high value in 1991 was probably due to migration of fish from Eastern Basin to Sätofta Basin (Figure 1) (See also Hamrin, 1993). In Eastern Basin, the catches were five times as large before than after fish reduction (Regression, Ln-transformed values, $F = 6.9$, $P = 0.059$) (Figure 1). Fish biomass was relatively constant after fish reduction (years 1990–1994). In Western Basin, the size of the catches were at least twice as large before as after the reduction, but there was no statistically significant decrease from 1987 to 1994 (Regression, Ln-transformed values, $F = 2.8$, $P = 0.165$) (Figure 1).

Although there are differences between the three basins, the general pattern is that the fish abundances increased in all basins. In Sätofta Basin, the abundance increased during the period 1987–1994 from ~3 000 to ~12 000 fish per 20 min of trawling (Regression, Ln-transformed values, $F = 8.9$, $P = 0.009$), due mainly to the very high abundance of young-of-the-year in 1992 and 1994 (Figure 2). The increase in young-of-the-year was not statistically significant (Regression, Ln-transformed values, $F = 3.1$, $P = 0.152$). In Eastern Basin, the fish abundance did not show any monotonic changes from 1987 to 1994 (Regression, Ln-transformed values, $F = 2.0$, $P = 0.229$). Young-of-the-year abundance however, increased during this period (Regression, Ln-transformed values, $F = 10.2$, $P = 0.033$) (Figure 2). Finally, in Western Basin the abundance

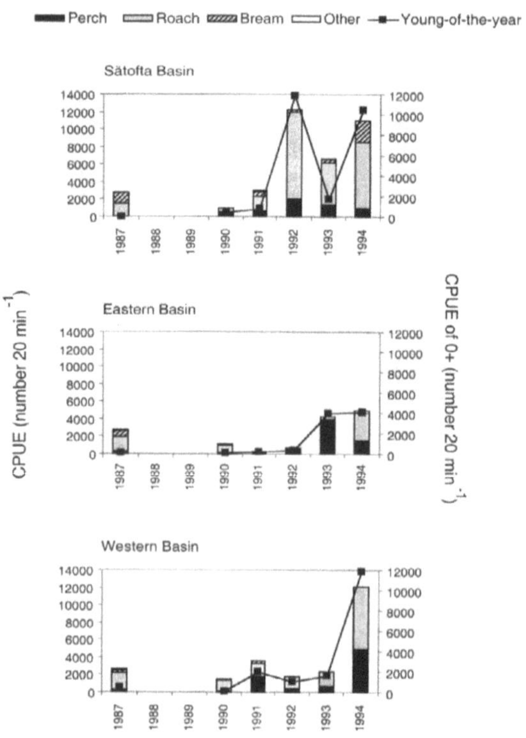

Figure 2. Fish abundance (CPUE, number 20 min^{-1} trawling) in different years in Sätofta Basin, Eastern Basin and Western Basin.

was higher in 1994 than in the other years, although the difference was not statistically significant (Regression, Ln-transformed values, $F = 2.8$, $P = 0.167$) (Figure 2). This was mainly due to very high abundances of young-of-the-year roach and perch in 1994. The abundance of the adult fish (>70 mm) decreased during the period 1987–1994 (Regression, Ln-transformed values, $F = 34.8$, $P = 0.004$).

Size- and year-class distribution

In all basins only a small proportion of the caught perch were large enough to be piscivorous. In Sätofta Basin and Eastern Basin the rather low amount of young-of-the-year perch in 1987 disappeared after the fish reduction, whereas the amount of young-of-the-year was high throughout the study-period in Western Basin (Figure 3). A strong year class of young-of-the-year perch one year did not seem to affect the biomass of the adult perch the following year (Figure 4).

The abundance of piscivorous zander, i.e. 1+ or larger, varied between years in all basins (Figure 3). In both Sätofta Basin and Eastern Basin, the abundances of zander was lower before the fish reduction than after the reduction, whereas the biomasses were high even before the fish reductions, indicating that the variations were due to an increase in the recruitment of young-of-the-year zanders after the cyprinid reductions (Figures 3 & 4). In Western Basin, the biomass of zander was similar throughout the investigation period, whereas the total abundance steadily decreased from 1987 to 1994 (Figures 3 & 4).

Although there was a decrease in the roach biomass from 1987 to 1994 in all basins (except 1991 in Sätofta Basin), recruitment of young-of-the-year roach was less explosive and took longer time than the perch population following the biomanipulation (Figures 3 & 4). In Sätofta Basin, high numbers of young-of-the-year roach were seen in 1992 and 1994, and in Western Basin in 1992–1994, but high numbers were found in Eastern Basin only by 1993 and 1994 (Figures 3 & 4).

In Sätofta Basin, the bream population showed rather high abundances of young-of-the-year, whereas adult bream dominated in the other basins. Both abundance and biomass of bream decreased in Eastern Basin and Western Basin after fish reduction, but there was no obvious change in size distribution. One exception is 1994 in Eastern Basin, when there was a strong young-of-the-year class (Figures 3 & 4). Sätofta Basin also showed a very high amount of young-of-the-year bream that year.

The relationship between abundance of young-of-the-year fish and abundance of piscivorous perch was plotted for Sätofta Basin and Eastern Basin, both for the entire basin and for the 3 m depth. This showed that only when the abundance of piscivorous perch was very low (<5 in Sätofta Basin, <1 in Eastern Basin) the abundance of young-of-the-year was high (Figure 5). This was true for both the 3 m depth and the whole lake in Sätofta Basin, but only at the 3 m depth in Eastern Basin.

Discussion

Several criteria need to be fulfilled if the removal of planktivores is to have a long-lasting, positive effect on the lake ecosystem. One criterion is that the total planktivorous fish biomass must be below a certain threshold, another is that recruitment of planktivores

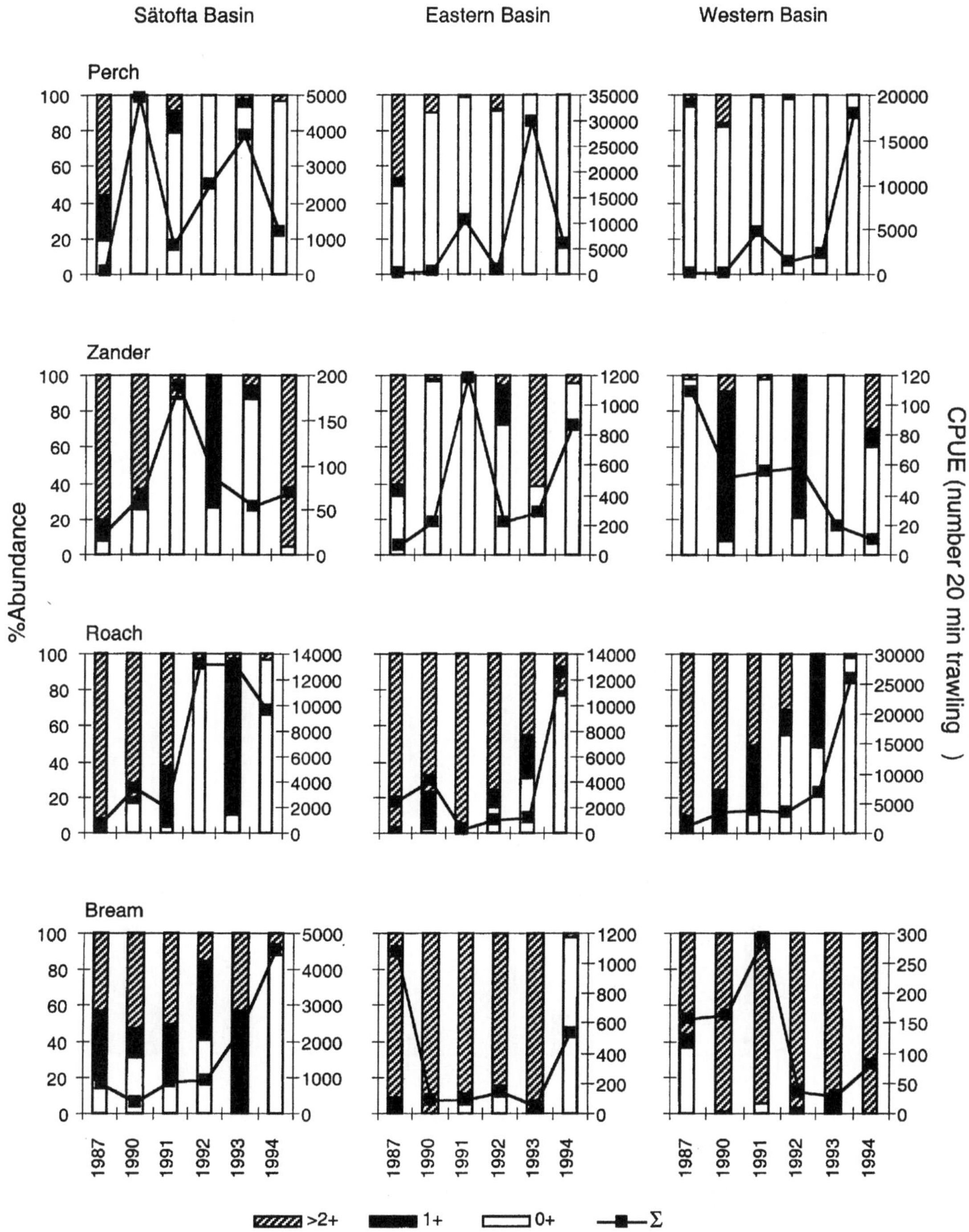

Figure 3. Proportions (based on abundances) of different year-classes and total biomass of perch, zander, roach and bream in different years in Sätofta Basin (left), Eastern Basin (middle) and Western Basin (right). Note the different scales on the y_2-axes.

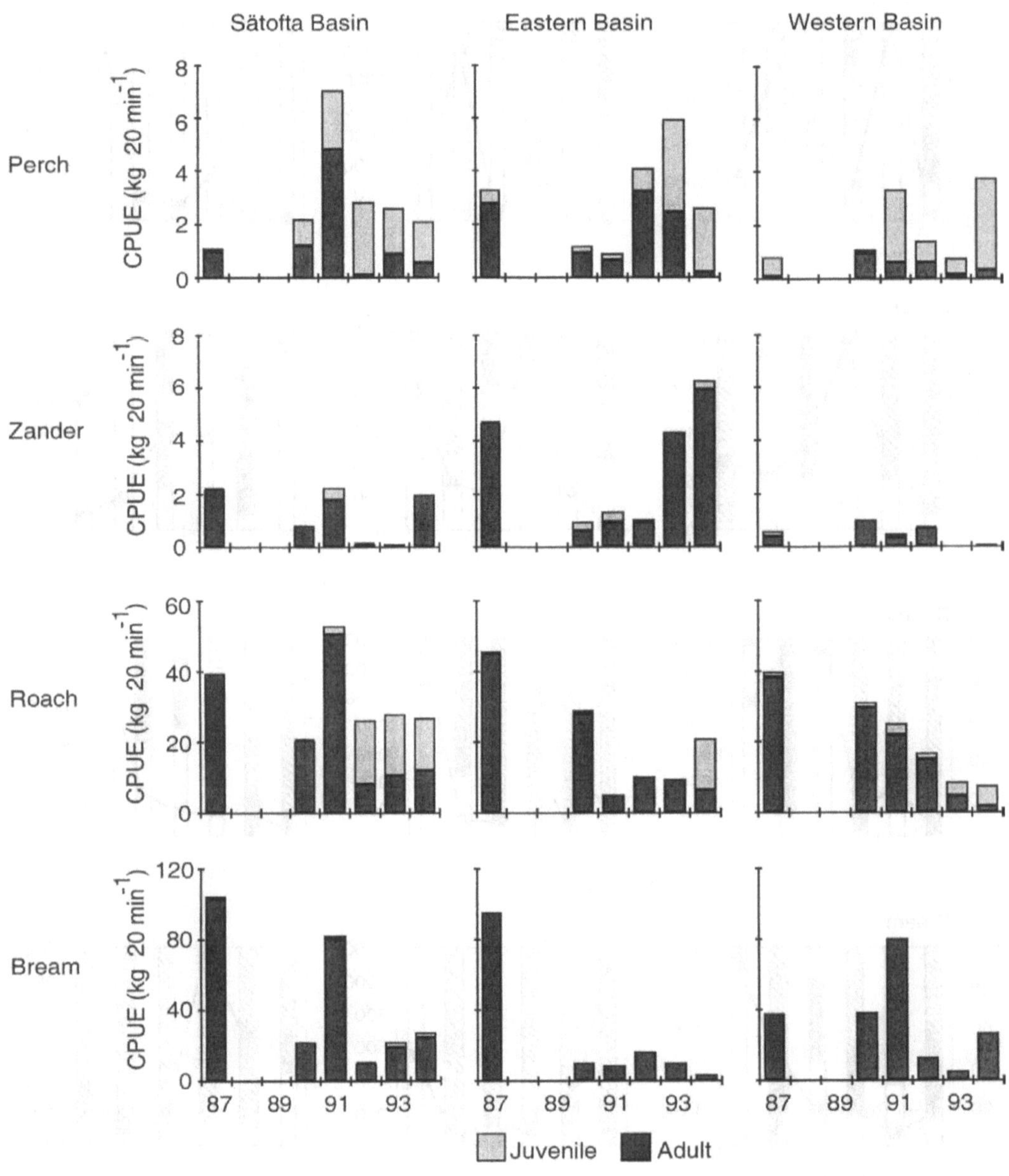

Figure 4. Biomass of juvenile and adult roach, bream, perch and zander in different years in Sätofta Basin (left), Eastern Basin (middle) and Western Basin (right). Note the different scale on the y-axes.

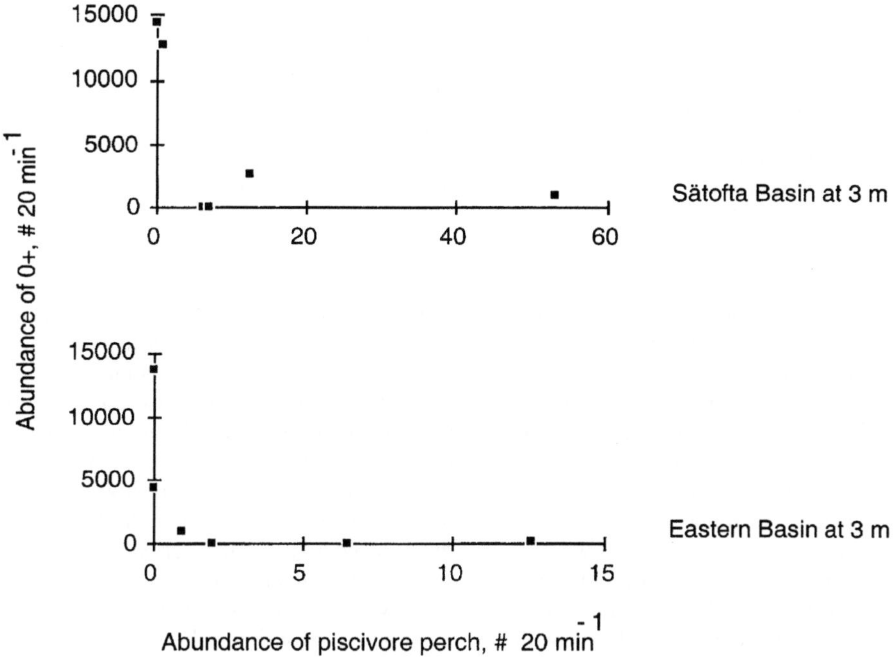

Figure 5. Relationship between abundance of young-of-the-year fish and piscivorous perch at the depth of 3 m (above) in Sätofta Basin and Eastern Basin (below).

to adult stages cannot be too large and a third criterion is that there must be a certain proportion between piscivores and planktivores. This implies that factors must exist which hold back the compensatory mechanisms which bring back the fish biomasses to the same level as they were before the biomanipulation. One such mechanism is the common explosion of young-of-the-year following a fish reduction and/or large immigrations from nearby lakes and running waters. Another, is that piscivores able to consume different sizes of fishes, e.g. both young-of-the-year and larger adult planktivores, must be present.

Fish biomass and size distribution

The critical fish biomass below which large herbivores such as *Daphnia* can dominate the zooplankton community, and above which these large herbivores become scarce, differs depending on which fish species dominate the fish community. Benndorf (1990) presents some critical fish biomasses which would permit *Daphnia* to survive: for young-of-the-year it is 20–50 kg ha^{-1}, for small fish species as *Leucaspius* 120 kg ha^{-1}, and for roach and bream 200 kg ha^{-1}. In Sätofta Basin and Eastern Basin, the fish biomasses were well above these levels before the fish reduction, but in Western Basin, the biomass of bream and roach was lower than this critical level (Benndorf, 1990) (Table 1). Total fish biomass decreased by 50–80% after fish reduction in all basins of Lake Ringsjön, which is similar to results reported from studies where planktivores were reduced (Jeppesen et al., 1990b; Søndergaard et al., 1990; Meijer et al., 1994; Horppila et al., 1996). The decrease was mainly due to a decrease in bream, but also roach decreased. The fish biomass has for the most part remained low in all basins of Lake Ringsjön after the fish reduction, which is 3–5 years for the different basins. Only in 1991 did the biomass in Sätofta Basin return to its prereduction level, which was attributed to migration from Eastern Basin. The increase of fish biomass in Sätofta Basin was approximately the same as the decrease in Eastern Basin. In 1992, the fish biomass returned to the 1990 level without any further attempts to reduce fish biomass in Sätofta Basin.

We found a strong increase in young-of-the-year after fish reduction, mainly of perch but later roach

Table 1. Fish biomass in the three basins of Lake Ringsjön before and after the fish reduction

	Before		After		Lake surface
	ton	kg ha^{-1}	ton	kg ha^{-1}	ha
Sätofta basin	159	376	59	139	423
Eastern basin	625	305	125	61	2050
Western basin	223	150	103	69	1484

also increased. Furthermore, there was a negative relationship between abundance of piscivorous perch and young-of-the-year fish in the top 3 m of both Sätofta Basin and Eastern Basin. This is in agreement with Benndorf (1995), who suggested that piscivorous perch may have been the main regulator of the abundance of young-of-the-year fish after biomanipulation in the reservoir Bautzen. He found that a decrease in abundance of piscivorous perch was followed by high abundance of young-of-the-year fish. This pattern seems to be influenced by the morphometry of the lake, because if we compare Sätofta Basin with Eastern Basin we found the same pattern at the 3 m depth and the whole basin in Sätofta Basin but not in Eastern Basin. The 3 m depth represents about 80% of the Sätofta Basin, whereas Eastern Basin is deeper and all the piscivorous perch were found at deeper depths than the young-of-the-year.

Trawling usually underestimates the biomass of larger piscivorous fish since they can escape the trawl better than other fish. This bias would be the same over the entire study if the probability that the fish saw the trawl was the same over time. In our study however, we had an increase in Secchi depth over time and a synchronous decrease in trawl catches. Although Secchi depth in Western Basin increased before the fish reduction in that basin, all basins showed the same pattern of a decrease in trawl catches and an increase of the young-of-the-year after the fish reduction. Furthermore, the number of young-of-the-year in the catches was not related to the mean size of the young-of-the-year (Figure 6). Thus, we do not think the catches in our test trawling fluctuated between years.

Recruitment of young-of-the-year to adults

Several studies have shown that there is an increase in the number of young-of-the-year fish after a fish reduction (e.g. Meijer et al., 1995; Horppila et al., 1996). If the biomass of planktivores is to remain low after the fish reduction, external P loading and recruitment of young-of-the-year planktivores to adult stages must be low. One way of controlling young-of-the-year is via a suitable piscivore, otherwise compensatory mechanisms will return the lake to its initial conditions (Benndorf, 1990). In Lake Ringsjön, abundances of young-of-the-year perch were high in all basins and of young-of-the-year roach in Sätofta Basin following the fish reduction. Although young-of-the-year constituted more than 50% of the biomass of the catch some years, there was no corresponding increase in adult perch or roach in following years, e.g. there was no increase in adult planktivore biomass after the fish reduction (Figures 3 & 4). Although this result is positive from the biomanipulation point of view, we do not know its cause. It may have been due to direct predation from piscivores, from juvenile competition and starvation during the winter, or some other unknown factor.

Proportion of piscivorous fish

One expectation on the reduction of the planktivorous fish was that the proportion of piscivores would increase in relation to the planktivorous fish (Hamrin et al., 1991, 1993). This would in turn both decrease the predation pressure on large grazing zooplankters and, if the top trophic level could be successful in regulating the zooplanktivorous fish, increase the predation pressure on smaller (often planktivorous) fish (Persson et al., 1991; Hamrin et al., 1991, 1993). Experiences from shallow Danish lakes showed that the ratio piscivores:planktivores should be 0.5 if the fish reduction is to have a lasting effect (Jeppesen pers. comm., see also Jeppesen et al., 1990a,b; Søndergaard et al., 1990). Benndorf (1990) suggested that 30–40% piscivores are necessary to stabilize a fish community. Benndorf (1990) argues further, that to achieve a long lasting effect of a biomanipulation, there must be a sufficient number of piscivores able to consume the

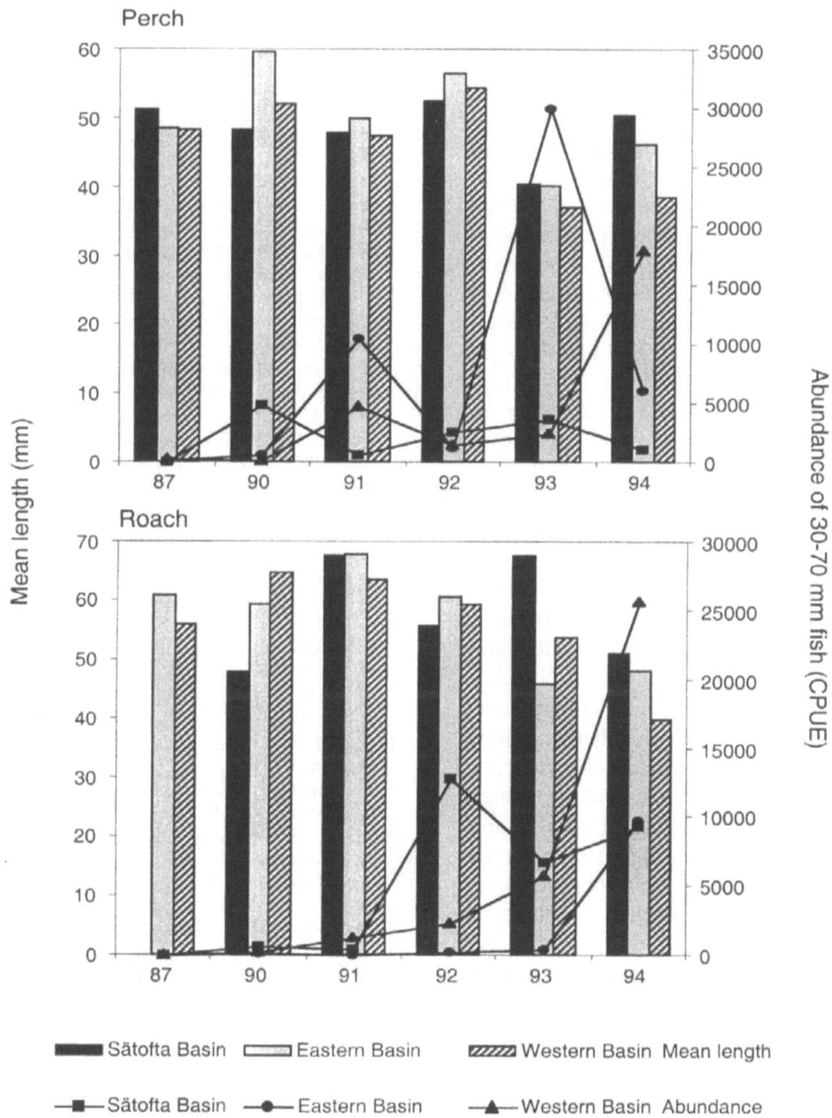

Figure 6. Mean length (mm) and abundance of fish between 30 and 70 mm (CPUE, number 20 min^{-1} trawling) of perch (above) and roach (below) in Sätofta, Eastern and Western Basins.

larger planktivores, as well as piscivores which consume young-of-the-year when they are abundant and switch to benthic prey rather than zooplankton when young-of-the-year are scarce.

The ratio of piscivores:planktivores increased from 12 to 40% in Eastern Basin, from 4 to 14% in Western Basin and in Sätofta Basin it stayed at 8% for the whole period. The fish reduction in Western Basin was not terminated until 1992, and since perch requires several years to become piscivorous (Collette et al., 1977), it is possible that the ratio will continue to increase. The final ratio is thus much lower than that suggested by Jeppesen (pers. comm.; Jeppesen et al., 1990a,b) and Benndorf (1990). Despite this low ratio piscivores:planktivores, the water quality has improved in all basins, which leads us to question the generality of this ratio. There are however, several differences between the Danish lakes and Lake Ringsjön which may influence the interpretation of this ratio. Lake Ringsjön is larger and deeper than the Danish lakes, which could be one factor affecting the pattern of a decreasing negative effect of planktivores on the zooplankton community with increasing depth of the lake (Wright & Shapiro, 1990; Jeppesen et al., 1990a; Shapiro, 1995). Thus, it is possible for a lower ratio of piscivores:planktivores to have the same effect in Lake Ringsjön as in the Danish lakes (Jeppesen et al., 1990a,b). Another difference between Lake Ringsjön and the lakes studied by Jeppesen et al. (1990a,b) is that the nutrient loading was lowered in Lake Ringsjön before the biomanipulation, which was not the case in the Danish lakes. In Lake Ringsjön, we actually have the best result in the deepest of the three basins, which also has the highest piscivore:planktivore ratio. However, only in Eastern Basin is the ratio at least close to that suggested by Jeppesen. It is also unclear how the low ratio in Sätofta Basin and Western Basin affect Eastern Basin since fish can migrate between the three basins.

In summary, there is still a much lower biomass of fish in all basins of Lake Ringsjön than before the fish reduction, and this is mainly due to a strong reduction of bream, but also of roach. The amount of piscivorous fish is very low in two of the three but considerably higher in the third basin, which ought to restrict zooplanktivory in this third basin for a long period of time. Despite the very low amount of piscivores in two of the basins, there was a continuous decrease of nutrients and phytoplankton biomass and an increase in Secchi depth which seem to last and demonstrate a positive result of the fish reduction.

References

Andersson, G. H., H. Berggren, G. Cronberg & C. I. Gelin, 1978. Effects of planktivorous fish on organisms and water chemistry in eutrophic lakes. Hydrobiologia 59: 9–15.

Benndorf, J., 1990. Conditions for effective biomanipulation; conclusions derived from whole-lake experiments in Europe. Hydrobiologia 200/201: 187–204.

Benndorf, J., 1995. Possibilities and limits for controlling eutrophication by biomanipulation. Int. Rev. Ges. Hydrobiologia 80: 519–534.

Brabrand, Å., B. A. Faafeng & J. P. Nilssen, 1986. Juvenile roach and invertebrate predators: delaying the recovery phase of eutrophic lakes by suppression of efficient filter feeders. J. Fish. Biol. 29: 99–106.

Brabrand, Å., B. A. Faafeng & J. P. M. Nilssen, 1990. Relative importance of phosphorus supply to phytoplankton production: Fish excretion versus external loading. Can. J. Fish Aquat. Sci. 47: 364–372.

Carpenter, S. R., J. F. Kitchell & J. R. Hodgson, 1985. Cascading trophic interactions and lake productivity. BioScience 35: 634–639.

Collette, B. B., M. A. Ali, K. E. F. Hokanson, M. Nagiec, S. A. Smirnov, A. H. Thorpe, A. H. Weatherley, & J. Willemsen, 1977. Biology of Percids. J. Fish. Res. Bd Can. 34: 1890–1899.

Hamrin, S. F., 1993. Lake restoration by cyprinid control in Sätofta Bay (Lake Ringsjön). Verh. Int. Ver. Limnol. 25: 487–493.

Hamrin, S. F., 1999. Planning and excecution of the fish reduction in Lake Ringsjön. Hydrobiologia 404: 59–63.

Hamrin, S. F., E. Bergstrand, G. Cronberg & P. Romare, 1991. Lake restoration by cyprinid reduction. The development of Lake Ringsjön since 1966 and effects of the fish reduction 1988–1990. Mimeographed. Department of Limnology, Lund. (In Swedish).

Hamrin, S. F., E. Bergstrand, G. Cronberg, A. Persson, P. Romare, & J. Strand, 1993. Lake restauration by cyprinid reduction. The development of Lake Ringsjön during the period 1987–1992. Sjörestaurering genom cyprinidreduktion. Ringsjöns utveckling under perioden 1987–1992. Mimeographed. Department of Limnology in Lund and Institute of Freshwater Research in Drottningholm. (In Swedish).

Hansson, L.-A., M. Enell & E. Bergman, 1999. Lake Ringsjön: its catchment area, its history and its importance. Hydrobiologia 404: 1–7.

Hartman, J. & W. Nümann, 1977. Percids in Lake Constance, a lake under eutrophication. J. Fish. Res. Bd. Can. 34: 1670–1677.

Henrikson, L., H. G. Nyman, H. G. Oscarson & J. A. E. Stenson, 1980. Trophic changes without changes in the external nutrient loading. Hydrobiologia 68: 257–263.

Horppila J. & T. Kairesalo, 1990. A fading recovery: the role of roach (*Rutilus rutilus* L.) in maintaining high phytoplankton productivity and biomass in Lake Vesijärvi, southern Finland. Hydrobiologia 200/201: 153–165.

Horppila, J., K. Nyberg, H. Peltonen & T. Turunen, 1996. Effects of five years of intensive trawling on a previously unexploited smelt stock. J. Fish Biol. 48: 329–340.

Jeppesen, E., J. P Jensen, P. Kristensen, M. Søndergaard, E. Mortensen, O. Sortkjær & K. Olrik, 1990a. Fish manipulation as a lake restoration tool in shallow, eutrophic temperate lakes 2: threshold levels, long-term stability and conclusions. Hydrobiologia 200/201: 219–227.

Jeppesen, E., M. Søndergaard, E. Mortensen, P. Kristensen, B. Riemann, H. J. Jensen, J. P. Müller, O. Sortkjær, J. P. Jensen, K. Christoffersen, S. Bosselmann & E. Dall, 1990b. Fish manipulation as a lake restoration tool in shallow, eutrophic temperate lakes 1: cross-analysis of three Danish case-studies. Hydrobiologia 200/201: 205–218.

Kitchell, J. F. (Ed.) 1992. Food Web Management. A case study of lake Mendota. Springer-Verlag, New York.

Lampert, W., W. Flecker, H. Rai & B. E. Taylor, 1986. Phytoplankton control by grazing zooplankton: A study on the spring clear-water phase. Limnol. Oceanogr. 31: 478–490.

Leopold, M., M. Bni'nska & W. Nowak, 1986. Commercial fish catches as an index of lake eutrophication. Arch. Hydrobiol. 106: 513–524.

Lessmark, O., 1983. Competition between perch (*Perca fluviatilis*) and roach (*Rutilus rutilus*) in south Swedish lakes. Dissertation. Institute of Limnology, University of Lund, Sweden. 172 pp.

Meijer, M.-L., E. Jeppesen, E. van Donk, B. Moss, M. Scheffer, E. H. R. R. Lammens, E. H. van Nes, B. Faafeng & J.

P. Jensen, 1994. Long term responses on fish stock reduction in small shallow lakes. Interpretation of five year results of four biomanipulation cases in The Netherlands and Denmark. Hydrobiologia 275/276: 457–466.

Meijer, M.-L., E. H. R. R. Lammens, A. J. P. Raat, J. G. P. Klein Breteler & M. P. Grimm, 1995. Development of fish communities in lakes after biomanipulation. Neth. J. Aquat. Ecol. 29: 91–101.

Oksanen, L., S. D. Fretwell, J. Arruda, & P. Niemelä, 1981. Exploitation ecosystems in gradients of primary productivity. Am Nat 118: 240–261.

Persson, A., 1997. Effects of fish predation and excretion on the configuration of aquatic food webs. Oikos 79: 137–146.

Persson, L., G. Andersson, S. F. Hamrin & L. Johansson, 1988. Predator regulation and primary production along the productivity gradient of temperate lake ecosystems. In S. R. Carpenter (ed.), Complex Interactions in Lake Communities. Springer-Verlag. New York: 45–65.

Persson, L., S. Diehl, L. Johansson, G. Andersson & S. F. Hamrin, 1991. Shifts in the fish communities along the productivity gradient of temperate lakes – patterns and the importance of size-structured interactions. J. Fish. Biol. 38: 281–293.

Riemann, B., K. Christoffersen, H. J. Jensen, J. P. Müller, C. Lindegaard & S. Bosselmann, 1990. Ecological consequences of a manual reduction of roach and bream in a eutrophic, temperate lake. Hydrobiologia 200/201: 241–250.

Schriever, P., J. Bøgestrand, E. Jeppesen & M. Søndergaard, 1995. Impact of submersed macrophytes on fish-zooplankton-phytoplankton interactions: large-scale enclosure experiments in a shallow eutrophic lake. Freshwat. Biol. 33: 255–270.

Shapiro, J., 1995. Lake Restoration by biomanipulation – a personal view. Envir. Rev. 3: 83–93.

Søndergaard, M., E. Jeppesen, E. Mortensen, E. Dall, P. Kristensen & O. Sortkjær, 1990. Phytoplankton biomass reduction after planktivorous fish reduction in a shallow, eutrophic lake: a combined effect of reduced internal P-loading and increased zooplankton grazing. Hydrobiologia 200/201: 220–240.

van Donk, E., M. P. Grimm, R. D. Gulati, P. G. M. Heuts, W. A. de Kloet & E. van Liere, 1990a. First attempt to apply whole-lake food-web manipulation on a large scale in The Netherlands. Hydrobiologia 200/201: 291–302.

van Donk, E., M. P. Grimm, R. D. Gulati & J. P. G. Klein Breteler, 1990b. Whole-lake food-web manipulation as a means to study community interactions in a small ecosystem. Hydrobiologia 200/201: 275–290

Wright, D. & J. Shapiro, 1990. Refuge availability: a key to understanding the summer disappearance of *Daphnia*. Freshwat. Biol. 24: 43–62.

Hydrobiologia **404:** 77–87, 1999.
L.-A. Hansson & E. Bergman (eds), Nutrient Reduction and Biomanipulation as Tools to Improve Water Quality: The Lake Ringsjön Story.

Lack of a top-down effect on the zooplankton community after a cyprinid reduction

Eva Bergman[1] & Eva Bergstrand[2]
[1]*Department of Limnology, Insitute of Ecology, Ecology Building, University of Lund, SE-223 62 Lund, Sweden*
[2]*Institute of Freshwater Research, SE-178 93 Drottningholm, Sweden*
Current address: Department of Biology, Karlstad University, S-651 88 Karlstad, Sweden

Key words: zooplankton community, cyprinid reduction, biomanipulation, top-down effect, indirect bottom-up effect

Abstract

Abundance, biomass and sizes of crustaceans were followed before, during and after a fish reduction in the three basins of Lake Ringsjön, southern Sweden. Zooplankton were sampled over the summer season (May–September) for eight years. Although the fish reduction occurred in different years in the three basins, the zooplankton community did not differ between basins. Between-year differences in biomass occurred in all basins, but these were not correlated with the time before and after the fish reduction, with the exception of the *Eubosmina* species. Analyses were performed on total biomass, as well as on biomass of *Daphnia galeata*, *D. cucullata*, *Bosmina longirostris*, *Eubosmina coregoni*, *E. crassicornis*, calanoid and cyclopoid copepods. Both calanoid copepods and *Daphnia cucullata* were larger before, than after the fish reduction. The same was true for a proportional length measurement of all grazers (=cladocerans+calanoid copepods). We found no evidence of a top-down effect on zooplankton biomass or length following the reduction in planktivorous fish. The possibility of an 'indirect bottom-up effect' due to cyprinid fish directly affecting nutrient availability is discussed.

Introduction

Top level predators having an effect on prey at lower trophic levels has been known since the sixties (Hrbacek et al., 1961; Brooks & Dodson, 1965). Several years later the term biomanipulation was coined by Shapiro et al. (1975) referring to their opinion that the top-down concept could be used in lake restoration. Since then much work has focused on the effect of planktivorous fish on the zooplankton community and further down the food chain. Several authors have reported significant changes in zooplankton community composition and sizes following a change in the abundance of zooplanktivorous fish based on both enclosure (Shapiro et al., 1975; Andersson et al., 1978) and full scale experiments (Shapiro & Wright, 1984; Sanni & Wærvågen, 1990; Lyche et al., 1990; Riemann et al., 1990; Søndergaard et al., 1990). McQueen et al. (1989) however, were critical to the mechanism behind the general pattern of a top-down effect from fish through zooplankton to phytoplankton and nutrients. They developed the concept of top-down versus bottom-up effects in order to distinguish biomanipulation from nutrient control in lake restoration approaches.

Recently, scientists have recognized that a simple pelagic food chain may not adequately describe trophic interactions in a lake. Interchanges between the littoral and the pelagial zone have been considered in terms of fish migration and the importance of submersed vegetation (Horppila et al., 1998). Furthermore, other effects of planktivorous fish (of which many are omnivorous and partly benthivorous) than their direct effect on zooplankton have also been recognized, such as effects on resuspension of sediments and detritus, effects on bioturbation, nutrient loss due to mortality, nutrient excretion and nutrient content of the removed fish (Drenner et al., 1986; Therkeld, 1987; Horppila & Kairesalo, 1990; Meijer et al., 1990; McQueen et al., 1992; Persson, 1997a,b). Thus, even if the top-down effect might work, several alternative mechanisms may be observed as a response to a large removal of planktivorous fish.

In the three basins of Lake Ringsjön, between 50 and 80% of the cyprinid biomass was removed during the period 1988–1992 in order to lower the predation pressure on the zooplankton and in that way increase the grazing on phytoplankton. We expected *Daphnia* to become more abundant and larger at the expense of smaller zooplankters such as *Chydorus* and *Bosmina* species. Zooplankton abundance and size were followed during the growth period (May–September) from 1987 to 1994 in all three basins.

Materials and methods

Study site

Lake Ringsjön is situated in southernmost Sweden (55° 52′ N; 13° 32′ O) and consists of three basins: Sätofta Basin (4.2 km^2, mean depth 3.0 m, maximum depth 17.0 m), Eastern Basin (20.5 km^2, mean depth 6.1 m, maximum depth 16.4 m) and Western Basin (14.8 km^2, mean depth 3.1 m, maximum depth 5.4 m) (Hansson et al., 1999). As with many other lakes in urban and agricultural areas, Lake Ringsjön became eutrophicated during the 1960s and 1970s. In the beginning of the 1980s, a *nutrient reduction programme* was initiated which reduced the external input of nutrients to the lake from phosphorus loadings exceeding 30 tons per year to around 10 tons per year. About ten years later (1989–1990), a *cyprinid fish reduction programme* (biomanipulation) was carried out in Sätofta Basin removing about 100 tons of fish (about 60% of the standing stock). During the winter of 1988, there was a fish-kill in Eastern Basin, eliminating about 500 tons of fish (about 80%) (Hamrin, 1999). The biomanipulation in Western Basin, which was performed in 1992, removed about 120 tons of fish (about 50%).

During the period 1987–1989, zooplankton sampling was carried out over the deepest part of the three basins using a plankton net (mesh size 150 μm, Ø=14 cm), which was drawn from the bottom to the surface. From 1990, the samplings were performed using a 1 m long plastic tube which could be closed in both ends. Samples were taken every second meter at eight depths in Sätofta Basin and Eastern Basin and at three depths in Western Basin. The sampled water was filtered through a plankton net (mesh size 150 μm) and preserved with formaldehyde to a final concentration of 4%. In 1990, parallel samplings using both the net and plastic tube method were taken in Eastern Basin in order to compare the methods. Analyses showed no systematic differences in the results using these two methods (Two-way ANOVA, P_{Method}=0.45, P_{Species}=0.0001, $P_{\mathrm{Method*Species}}$=1.0).

The animals were identified, counted and the length was measured. The biomass was calculated using the length-weight regressions of Bottrell et al. (1976). The genus *Bosmina* was divided into the species *Eubosmina coregoni thersites*, here named *E. coregoni*, *E. crassicornis* and *Bosmina longirostris* according to Lieder (1983a,b). The *Daphnia* complex which in an earlier investigation of the fish and zooplankton communities of Lake Ringsjön (Bergstrand, 1990) was identified as *Daphnia longispina s. l.*, has been specified as the *Daphnia galeata* complex, including *D. galeata* and *D. galeata x D. cucullata* according to Flössner & Kraus (1986). Here termed *D. galeata* only.

Statistics

To determine whether the composition of the zooplankton community changed after fish reduction, we performed a discriminant analysis (DA) of all species in the samples from 1987 to 1994, before and after the fish reduction. Since the fish reduction occurred in different years in the various basins, the number of years before and after the fish reduction varied (1990, 1988, 1992 in Sätofta Basin, Eastern Basin and Western Basin, respectively). Both biomass and abundance were analyzed. We also analyzed changes in biomass, abundance and length by means of variance and *t*-test analyses. We analyzed the length of the zooplankton in two different ways. First the three largest zooplankters were analyzed separately, and then we calculated a 'zooplankton community length' (ZCL). The ZCL was simply calculated by weighing the length of each species to its abundance and then calculating a mean for different species combinations such as daphnids, cladocerans and all grazers.

Results

Although the fish reductions occurred in different years in the three basins, a one-way ANOVA test showed no difference between the basins (1-way ANOVA $F_{2,21}$=1.404, P=0.268) (Figure 1). However, there was a significant between-year effect on the total zooplankton abundance (1-way ANOVA $F_{7,16}$=4.95, P=0.004). When the years were grouped as before and after the fish reduction in the different basins, a *t*-test

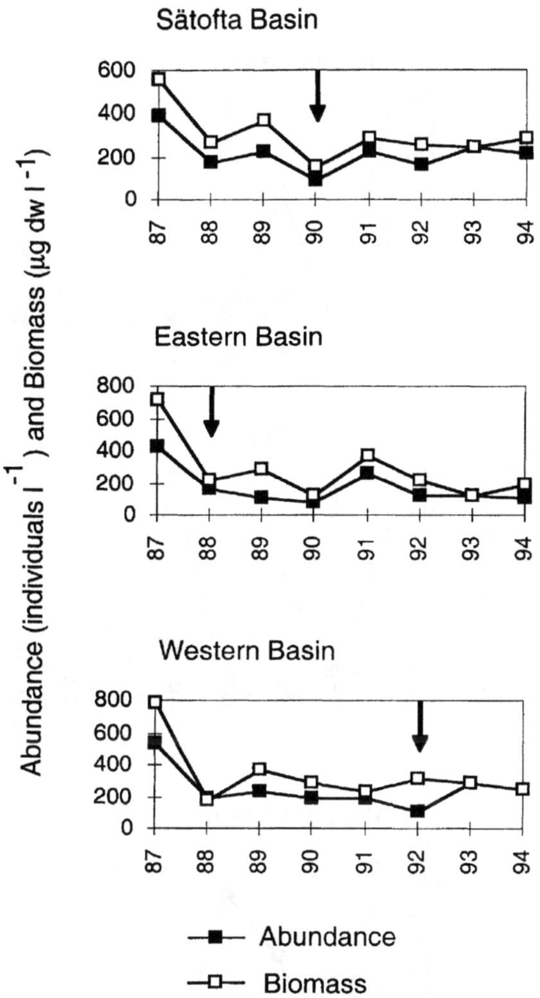

Figure 1. Abundance (summer means, ind. l^{-1}) and biomass (summer means, μg dw l^{-1}) of the crustacean community in Sätofta Basin, Eastern Basin and Western Basin from 1987 to 1994. Note the different axes for the three basins. Arrows indicate the years of fish reduction.

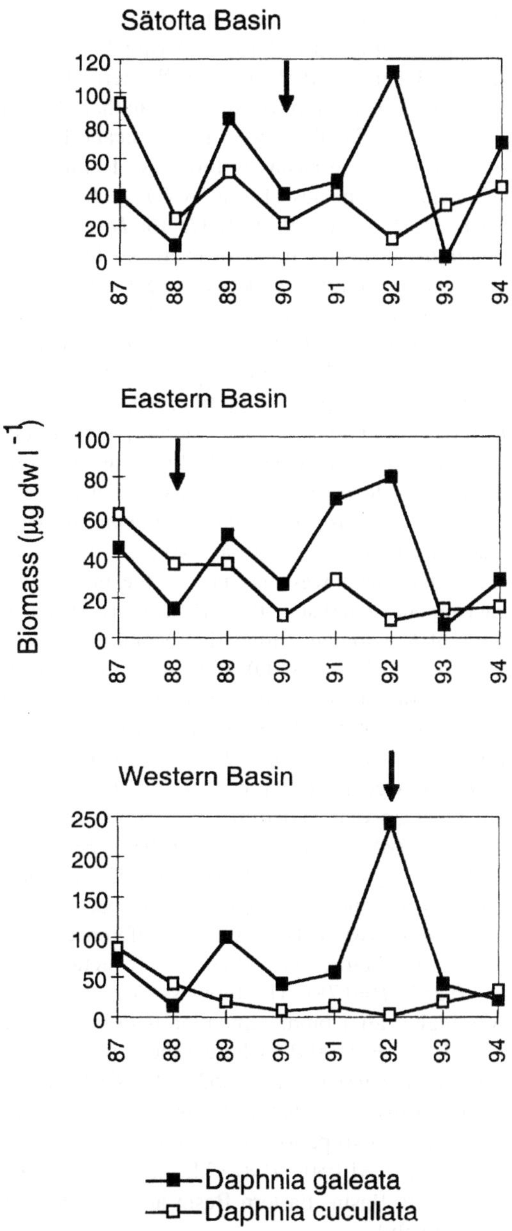

Figure 2. Biomass (summer means, μg dw l^{-1}) of *Daphnia galeata* and *D. cucullata* in Sätofta Basin, Eastern Basin and Western Basin from 1987 to 1994. Note the different axes for the three basins. Arrows indicate the years of fish reduction.

showed no difference in total abundance (t_{22}=1.694, P=0.104). The same analysis based on zooplankton biomass showed no difference between the basins (1-way ANOVA $F_{2,21}$=0.583, P=0.567) (Figure 1). The between-year variation in zooplankton biomass was significant (1-way ANOVA $F_{7,16}$=6.476, P=0.001), but there was no difference between the years before and after fish reduction (t_{22}=1.707, P=0.102).

The biomass (summer means) of *Daphnia galeata* fluctuated between 0 and 100 μg l^{-1}, except in 1992 in Western Basin where the biomass reached 250 μg

l^{-1}. The biomass of *D. cucullata* was high in 1987 (>60 μg l^{-1}) and varied between 4 and 50 μg l^{-1} the other years (Figure 2). In 1987 and 1988, the biomass of *Daphnia cucullata* was higher than the biomass of *D. galeata* in all three basins. During 1989–1992 in Sätofta Basin and 1989–1993 in Western Basin, the biomass of *D. galeata* was higher than that of *D. cucullata* (Figure 2). There was no significant difference between the three basins in biomass of either of the two *Daphnia* species (1-way ANOVA $F_{2,21}$=0.613, P=0.551, $F_{2,21}$=1.029, P=0.375 for *D. galeata* and *cucullata*, respectively). However, for both species there was a significant between-year difference (1-way ANOVA, $F_{2,21}$=6.304, P=0.001, $F_{2,21}$=7.600, P=0.0001 for *D. galeata* and *cucullata*, respectively), but no difference between the period before and after fish reduction (t-test, t_{22}=-0.101, P=0.920, t_{22}=1.878, P=0.074 for *D. galeata* and *cucullata*, respectively).

Bosmina longirostris reached the highest biomass of the three *Bosmina* species, varying between 0 and 160 μg l^{-1}. Also *Eubosmina coregoni* sometimes reached a biomass as high as 140 μg l^{-1}. *E. crassicornis* had the lowest biomass of the three species, varying between 0 and 50 μg l^{-1} (Figure 3). The biomass of *E. coregoni* varied significantly between the three basins (1-way ANOVA, $F_{2,21}$=3.789, P=0.039), whereas the other two *Bosmina* species (1-way ANOVA, $F_{2,21}$=0.956, P=0.400, $F_{2,21}$=0.514, P=0.605 for *E. crassicornis* and *B. longirostris*, respectively) showed no difference. There was a significant between-year variation for all three species (1-way ANOVA, $F_{7,16}$=2.944, P=0.035, $F_{7,16}$=6.665, P=0.001, $F_{2,21}$=7.226, P=0.001 for *E. coregoni*, *E. crassicornis* and *B. longirostris*, respectively). As for *B. longirostris* there was no difference between the period before and after fish reduction (t-test, t_{22}=0.283, P=0.78). The biomass of *E. crassicornis* was higher after biomanipulation than before (t-test, t_{22}=-5.192, P=0.0001), whereas the opposite was true for *E. coregoni* (t-test, t_{22}=2.537, P=0.019). However, if the basins were analyzed separately no difference between the two periods was found (t-test, t_6=2.261, P=0.064, t_6=1.050, P=0.334, t_6=1.241, P=0.261 for Sätofta Basin, Eastern Basin and Western Basin, respectively).

The biomasses of both calanoid and cyclopoid copepods were high in 1987 and then decreased, varying between 20 and 300 μg l^{-1} for calanoid copepods and between 0 and 225 μg l^{-1} for cyclopoid copepods (Figure 4). A one-way ANOVA test showed no significant effect of lake basin on the bio-

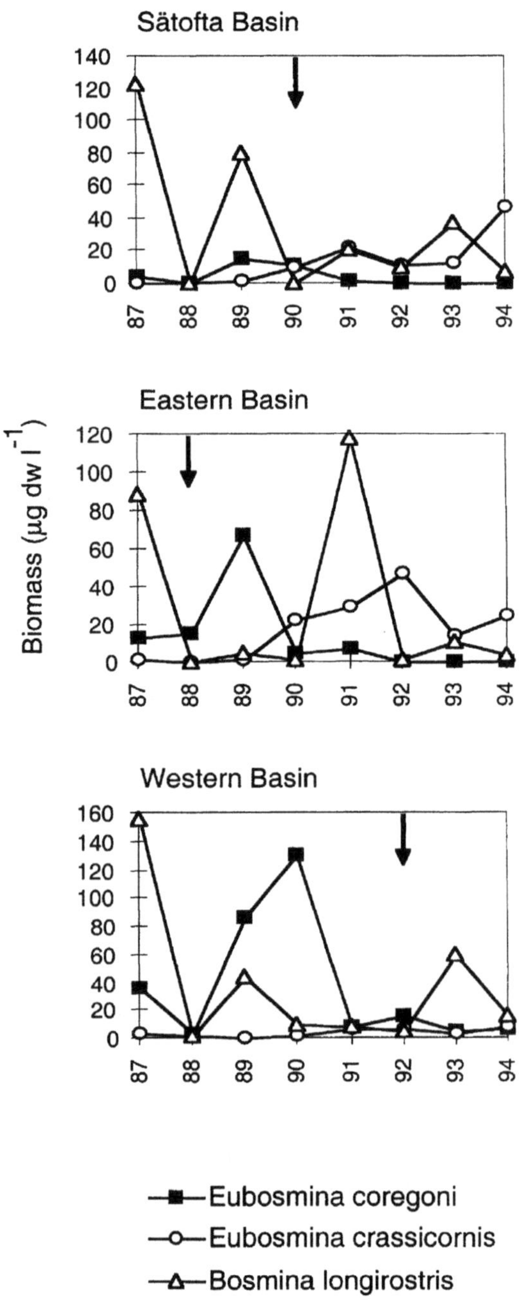

Figure 3. Biomass (summer means, μg dw l^{-1}) of *Eubosmina coregoni*, *E. crassicornis* and *Bosmina longirostris* in Sätofta Basin, Eastern Basin and Western Basin from 1987 to 1994. Note the different axes for the three basins. Arrows indicate the years of fish reduction.

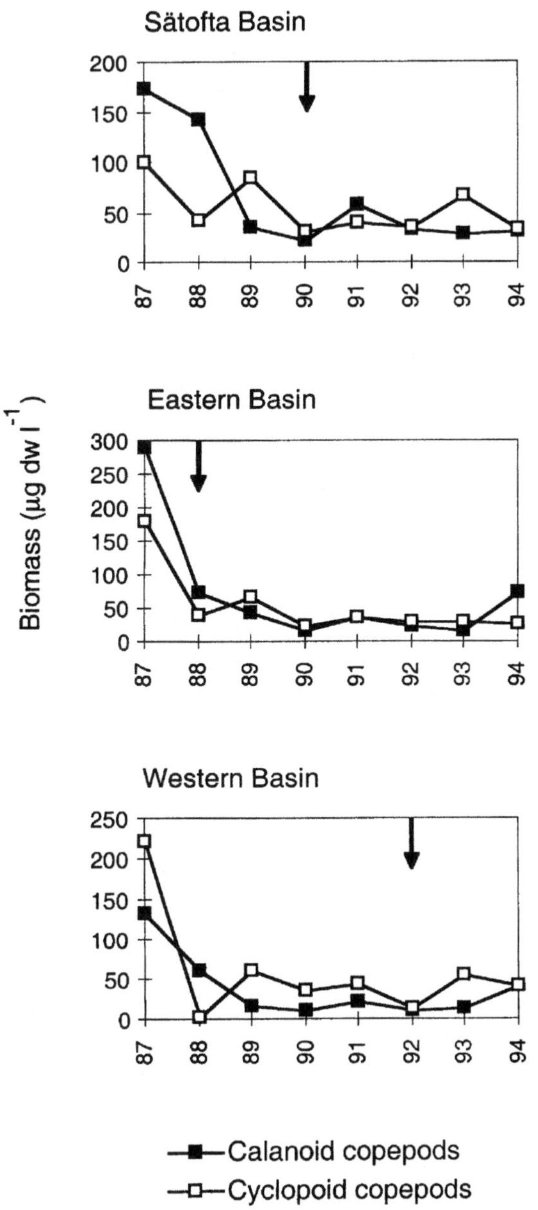

Figure 4. Biomass (summer means, μg dw l^{-1}) of calanoid and cyclopoid copepods in Sätofta Basin, Eastern Basin and Western Basin from 1987 to 1994. Note the different axes for the three basins. Arrows indicate the years of fish reduction.

Table 1. Three-way ANOVA testing the effect of month, treatment (=before and after fish reduction) and basin (Sätofta, Eastern or Western Lake Ringsjön) on the length (mm) of *Daphnia galeata, D. cucullata* and calanoid copepods

	SS	*df*	MS	*F*-value	*P*
Daphnia galeata					
Month	0.080	4	0.020	4.387	0.003
Treatment	0.011	1	0.011	2.388	0.126
Basin	0.030	2	0.015	3.247	0.044
Month*Treatment	0.042	4	0.010	2.285	0.067
Month*Basin	0.030	8	0.004	0.834	0.575
Basin*Treatment	0.016	2	0.008	1.812	0.170
Month*Basin*Treatment	0.046	8	0.006	1.256	0.278
Error	0.368	81	0.005		
Daphnia cucullata					
Month	0.040	4	0.010	3.220	0.016
Treatment	0.020	1	0.020	6.562	0.012
Basin	0.042	2	0.021	6.757	0.002
Month* Treatment	0.010	4	0.003	0.820	0.516
Month*Basin	0.015	8	0.002	0.620	0.759
Basin*Treatment	0.007	2	0.004	1.175	0.314
Month*Basin*Treatment	0.019	8	0.002	0.775	0.626
Error	0.265	86	0.003		
Calanoid copepods					
Month	0.041	4	0.010	3.288	0.015
Treatment	0.015	1	0.015	4.934	0.029
Basin	0.094	2	0.047	14.951	0.0001
Month*Treatment	0.003	4	0.001	0.270	0.897
Month*Basin	0.039	8	0.005	1.540	0.155
Basin*Treatment	0.000	2	0.000	0.033	0.967
Month*Basin*Treatment	0.021	8	0.003	0.857	0.556
Error	0.282	90	0.003		

mass of either of the two copepods ($F_{2,21}$=1.168, P=0.33, $F_{2,21}$=0.309, P=0.737 for calanoid and cyclopoid copepods, respectively). There was however, a significant between-year variation for both copepod species (1-way ANOVA, $F_{7,16}$=10.618, P=0.0001, $F_{7,16}$=3.949, P=0.011 for calanoid and cyclopod copepods, respectively). No differences in copepod biomass were found before and after fish reduction (t-test, t_{22}=1.903, P=0.07, t_{22}=1.125, P=0.273 for calanoid and cyclopoid copepods, respectively.

The length of *Daphnia galeata* and *D. cucullata* varied between 1.00–1.34 mm and 0.78–1.06 mm in Sätofta Basin, between 0.98–1.28 mm and 0.77–

Table 2. Three-way ANOVA testing the effect of month, treatment (=before and after fish reduction) and basin (Sätofta, Eastern or Western Lake Ringsjön) on the length (mm) of daphnids, cladocerans and all grazers (cladocerans+calanoid copepods)

	SS	*df*	MS	*F*-value	*P*
Daphnids					
Month	0.173	4	0.043	4.561	0.002
Treatment	0.005	1	0.005	0.491	0.485
Basin	0.040	2	0.020	2.080	0.131
Month*Treatment	0.030	4	0.007	0.779	0.542
Month*Basin	0.049	8	0.006	0.644	0.739
Basin*Treatment	0.001	2	0.000	0.047	0.954
Month*Basin*Treatment	0.021	8	0.003	0.279	0.971
Error	0.856	90	0.010		
Cladocerans					
Month	0.224	4	0.056	4.945	0.001
Treatment	0.008	1	0.008	0.699	0.405
Basin	0.072	2	0.036	3.163	0.047
Month*Treatment	0.109	4	0.027	2.409	0.055
Month*Basin	0.118	8	0.015	1.296	0.256
Basin*Treatment	0.011	2	0.006	0.501	0.607
Month*Basin*Treatment	0.048	8	0.006	0.531	0.830
Error	1.021	90	0.011		
All grazers					
Month	0.259	4	0.065	5.667	0.0001
Treatment	0.051	1	0.051	4.476	0.037
Basin	0.188	2	0.094	8.223	0.001
Month*Treatment	0.107	4	0.027	2.340	0.061
Month*Basin	0.144	8	0.018	1.574	0.144
Basin*Treatment	0.024	2	0.012	1.066	0.349
Month*Basin*Treatment	0.050	8	0.006	0.548	0.817
Error	1.026	90	0.011		

1.12 mm in Eastern Basin, and between 0.91–1.12 mm and 0.67–0.95 mm in Western Basin. The length of calanoid copepods varied between 0.91–1.07 mm, 0.79–1.00 mm and 0.78–0.90 mm in Sätofta Basin, Eastern Basin and Western Basin, respectively. There was an effect of month, basin and treatment (before and after fish reduction) on both *Daphnia cucullata* and calanoid copepods and of month and lake on *D. galeata* (Table 1). Both *Daphnia* species and the calanoid copepods were larger before than after fish reduction, except in Western Basin where *D. galeata* had the same length before and after. All these three species were longest in Sätofta Basin, smallest in Western Basin and of intermediate length in Eastern Basin (Figures 5, 6). The ZCL for the *Daphnia* species showed an effect of month, ZCL of cladocerans showed an effect of month and lake, and that of all grazers showed an effect of month, lake and treatment (before and after fish reduction) (Table 2). The daphnids were longer in May and June than during the rest of the summer, while cladocerans and all grazers were longer in June than in the other months, and were smaller in Western Basin than in the other basins. The ZCL for all grazers was lower following fish reduction (Figures 5, 6).

The biomass and abundance (summer means) of *Diaphanosoma brachyurum, Daphnia galeata, D. cucullata, Eubosmina crassicornis, E. coregoni thersites, Bosmina longirostris, Chydorus sphaericus, Leptodora kindtii*, calanoid copepods and cyclopoid copepods were analyzed using discriminant analysis (DA). DA discriminated 'before' and 'after' fish reduction in the three basins in 88% of the cases, but the multivariate test was not statistically significant ($F_{10,13}$=1.177, P=0.384 using biomass, $F_{10,13}$=1.438, P=0.265 using abundance). The biomass and abundance of *Eubosmina crassicornis* (increased following fish reduction) and *E. coregoni thersites* (decreased following fish reduction) were best at separating the two periods (i.e. had the highest loadings). The two potentially larger species, *Daphnia galeata* and *D. cucullata*, were not important for separation of the two periods.

Discussion

Numerous studies have shown how the abundance and length of, particularly *Daphnia*, species increase following removal of planktivorous fish from a lake (Brooks & Dodson, 1965; Sanni & Wærvågen, 1990; Lyche et al., 1990; Riemann et al., 1990; Søndergaard et al., 1990). These observations are in accordance with the different theories of top-down control of planktivore fish on the zooplankton community (Hrbacek et al., 1961; Brooks & Dodson 1965; Carpenter et al., 1985; Persson et al., 1988). In Lake Ringsjön, the only species demonstrating change after fish reduction was *Eubosmina coregoni* (a decrease) and *E. crassicornis* (an increase). These species also had the highest loadings in the discriminant analyses (DA). *Daphnia* species and calanoid copepods all showed between-year fluctuations, but these were not related to fish reduction. Hence, fluctuations in the zooplankton community could not support any top-down effect from the fish reduction.

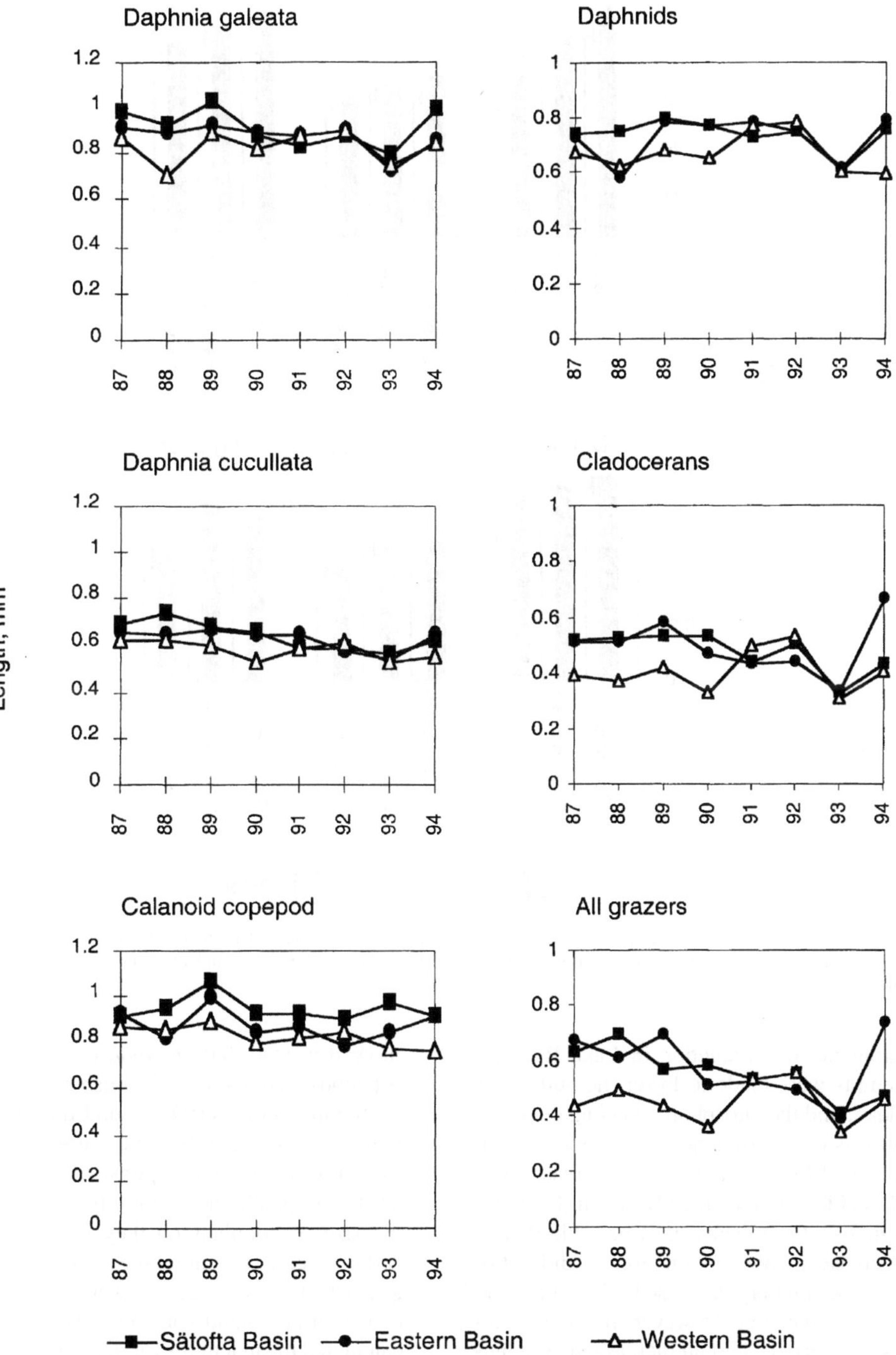

Figure 5. Length (mm) of *Daphnia galeata*, *D. cucullata* and calanoid copepods and ZCL (zooplankton community length, mm) of daphnids, cladocerans and all grazers in Sätofta Basin, Eastern Basin and Western Basin from 1987 to 1994.

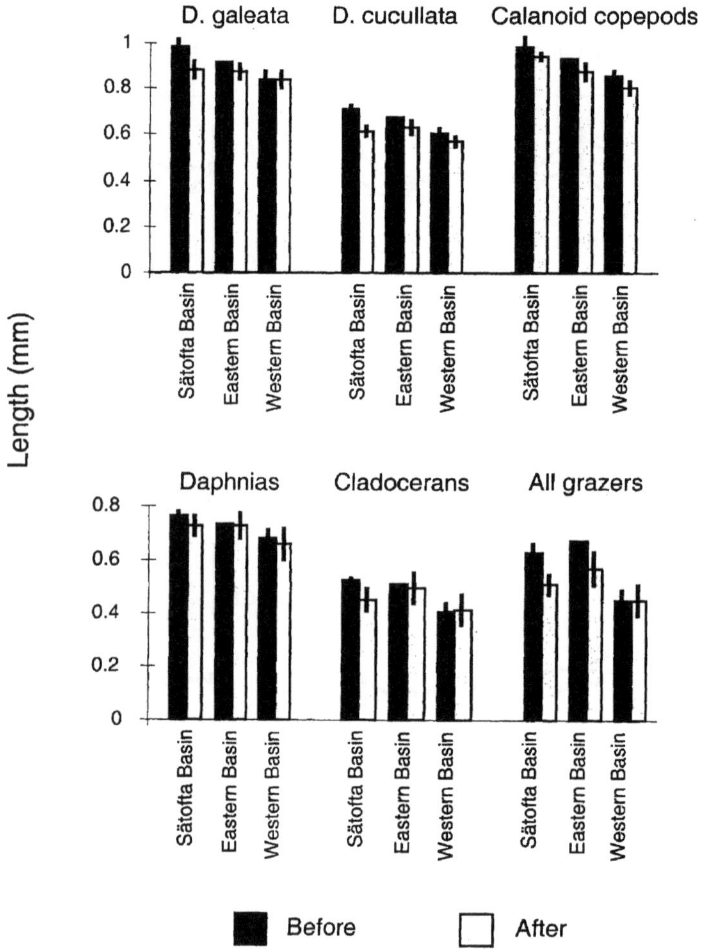

Figure 6. Length (mm ± SD) of *Daphnia galeata, D. cucullata* and calanoid copepods and ZCL (zooplankton community length, mm ± SD) of daphnids, cladocerans and all grazers in Sätofta Basin, Eastern Basin and Western Basin before and after fish reduction. All lengths are based on summer means.

Preceeding the fish reduction in Lake Ringsjön, several attempts were made to lower the nutrient input to the lake, and the phosphorus loading decreased from 20–30 tons yr^{-1} during the late seventies to about 10 tons yr^{-1} in 1985. At that time a decrease in the phytoplankton biomass had already begun (Cronberg, 1999; Bergman, 1997). Thus, it is possible that the expected increase in zooplankton biomass did not occur due to a contemporary decrease in their main food resource – phytoplankton. However, if this was true we would still expect to find an increase in zooplankton size due to the lowered planktivory following fish reduction. Surprisingly, however, the length analyses showed instead that *Daphnia cucullata* and calanoid copepods were larger before fish reduction than after.

Soranno et al. (1993) found that the zooplankton biomass was a complex and unpredictable variable to use for measuring the response of vertebrate predation on the zooplankton community. When they analyzed the size of zooplankton they did not use length of certain species but rather the mean length of cladocerans, which they suggested were a sensitive indicator of vertebrate planktivory. In Lake Ringsjön, we also calculated mean lengths for daphnids, cladocerans and cladocerans+calanoid copepods. The last group, however, was the only group to differ before and after fish reduction by being longer before the fish reduction than after. Thus, both biomass and length response failed to show any indications of a top-down effect following fish reduction in Lake Ringsjön. Similar results have been reported from Finland (Kairesalo et al., 1997; Horppila et al., 1998).

One reason for the lack of a clear top-down effect in Lake Ringsjön could be the increase in 0+ fish after fish reduction. Although the fish biomass in all basins remained considerably lower after fish reduction than before, there were high numbers in 0+ fish from one to five years after fish reduction (first perch but lately also roach) (Romare & Bergman, 1999; Bergman et al., 1999). A rough estimate of the predation pressure from 0+ fish (see Romare & Bergman, 1999) showed an actual increase following fish reduction. The effect of 0+ fish predation could also explain the larger size of both daphnids and cladocerans in May and June compared to later in the season (Figure 6). Thus, even if the 0+ fish do not survive the winter they could, due to a temporarily high predation pressure during late summer (July–September), stop the expected increase in zooplankton abundance and break the trophic cascade.

Despite the lack of a top-down effect at the zooplankton level in Lake Ringsjön, the fish reduction seemed to affect the lower trophic levels so that the phosphorus concentration and phytoplankton biomass continued to decrease and Secchi depth increased. The lake also seems to be more tolerant to high phosphorus loadings during years with high run-off (Bergman, 1999). A question arises: Do we have an indirect bottom-up effect instead of the expected top-down effect? Some other authors have suggested that changes in the amount of resuspended sediment and detritus, bioturbation of sediments and recycling of nutrients were the main factors linking cyprinid abundance to water quality (Andersson et al. 1978; Horppila & Kairesalo, 1990; Meijer et al., 1990; Brabrand et al., 1990). Later, Horppila et al. (1998) discussed the possibility that benthivorous and omnivorous fish may 'overtake' the top-down cascade and directly affect nutrient availability, for example by nutrient excretion, bioturbation and the effects of fish mortality. These direct couplings from cyprinids to the nutrient level improve water quality without involving the zoo- and phyto-plankton levels. Instead phytoplankton can show a decrease due to lower nutrient levels caused by the significant removal of fish.

Discussions about the importance of the littoral zone in relation to the pelagial zone are also relevant in this context, and include, for example, the distribution and density of macrophytes and migration of fish between pelagial and littoral habitats. Macrophytes create a refuge from fish predation for zooplankton (Schriver et al., 1995; Stansfield et al., 1997), but also compete with algae for nutrients (Moss, 1990) and favour the occurrence of pike (*Esox lucius*) (Benndorf 1990; Grimm & Backx, 1990). In Lake Ringsjön, the macrophyte stand is not very dense (Strand, this volume), so we do not expect zooplankton to find a predation-free refuge there. Migrating fish could translocate nutrients from the littoral to the pelagic zone, and much of the unexplained variability in lakes may be due to processes occurring on short time and small spatial scales (Kitchell et al., 1994 in Horppila et al., 1998). We need therefore to understand more about the timing and location of excreted nutrients (Horppila et al., 1998). In Lake Ringsjön, we do not have any measurements of migrating fish and no data on nutrient excretion. However, the littoral is closely linked to the pelagic system by fish foraging on benthic macroinvertebrates (Svensson et al., 1999). We must conclude that based on available data, it is not possible to explain why the zooplankton community did not respond to the reduction in cyprinid fish abundance in Lake Ringsjön. However, among the many possible causes discussed above, the most likely explanation is that the predation pressure on zooplankton was high even after fish reduction due to the considerable increase in 0+ fish abundance (Romare & Bergman, 1999). This notion is strengthened by the considerable increase in benthic macrofauna (Svensson et al., 1999), a food resource utilized by adult, but not by 0+ fish.

Acknowledgements

We thank two anonymous referees for valuable comments. The studies were funded by Swedish National Board of Fisheries, Swedish National Enviromental Protection Board, Sydvatten AB, The County Board of Skåne and the local authorities of Eslöv, Höör and Hörby.

References

Andersson, G., H. Berggren, G. Cronberg & C. I. Gelin, 1978. Effects of planktivorous fish on organisms and water chemistry in eutrophic lakes. Hydrobiologia 59: 9–15.

Benndorf, J., 1990. Conditions for effective biomanipulation; conclusions derived from whole–lake experiments in Europe. Hydrobiologia 200/201: 187–204.

Bergman, E., 1997. Lake Ringsjön in Scania, restoration by means of cyprinid reduction – the effect of the fish reduction. Final report. Department of Limnology. Lund University. Lund 1997. Mimeographed (in Swedish).

Bergman, E., 1999. Changes in the nutrient load and lake water chemistry in Lake Ringsjön, southern Sweden, from 1966 to 1996. Hydrobiologia 404: 9-18.

Bergman, E., S. F. Hamrin & P. Romare, 1999. The effects of cyprinid reduction on the fish community. Hydrobiologia 404: 65–75.

Bergstrand, E., 1990. Changes in the fish and zooplankton communities of Ringsjön, a Swedish lake undergoing man-made eutrophication. Hydrobiologia 191: 57–66.

Bottrell, H. H., A. Duncan, Z. M. Gliwicz, E. Grygierek, A. Herzig, A. Hillbricht-Illowska, H. Kurasawa, P. Larsson & T. Weglenska, 1976. A review of some problems in zooplankton productions studies. Norw. J. Zool 24: 419–456.

Brabrand, Å., B. A. Faafeng & J. P. M. Nilssen, 1990. Relative importance of phosphorus supply to phytoplankton production: Fish excretion versus external loading. Can. J. Fish Aquat. Sci. 47: 364–372.

Brooks, J. L. & S. I. Dodson, 1965. Predation, body size and the composition of plankton. Science 150: 28–35.

Carpenter, S. R., J. F. Kitchell & J. R. Hodgson, 1985. Cascading trophic interactions and lake productivity. BioScience 35: 634–639.

Cronberg, G., 1999. Qualitative and quantitative investigations of phytoplankton in Lake Ringsjön. Hydrobiologia 404: 27–40.

Drenner, R. W., S. T. Therkeld & M. D. McCracken, 1986. Experimental analysis of the direct and indirect effects of an omnivorous filter-feeding clupeid on plankton community structure. Can. J. Fish. aquat. Sci. 43: 1935–1945.

Flössner, D. & K. Kraus, 1986. On the taxonomy of the *Daphnia hyalina-galeata* complex (Crustacea: Cladocera). Hydrobiologia 137: 97–115.

Grimm, M. P. & J. J. G. M. Backx, 1990. The restoration of shallow eutrophic lakes, and the role of northern pike, aquatic vegetation and nutrient concentration. Hydrobiologia 200/201: 557–566.

Hamrin, S. F., 1999. Planning and execution of the fish reduction in Lake Ringsjön. Hydrobiologia 404: 59–63.

Hansson, L.-A., M. Enell & E. Bergman, 1999. Lake Ringsjön: its catchment area, its history and its importance. Hydrobiologia 404: 1–7.

Horppila, J. & T. Kairesalo, 1990. A fading recovery: the role of roach (*Rutilus rutilus* L.) in maintaining high phytoplankton productivity and biomass in Lake Vesijärvi, southern Finland. Hydrobiologia 200/201: 153–165.

Horppila, J., H. Peltonen, T. Malinen, E. Luokkanen & T. Kairesalo. 1998. Top-down or bottom-up effects by fish – issues of concern in biomanipulation of lakes. Restoration Ecology 6: 1–10.

Hrbacek, J., M. Dvorakova, V. Korinek & L. Prochazkova, 1961. Demonstrations of the effects of fish stock on the species composition and the intensity of the metabolism of the whole plankton association. Ver. Int. ver. Limnol. 14: 192–195.

Kairesalo, T., J. Keto & J. Horppila, 1997. Successful biomanipulation of Lake Vesijärvi: have cyanobacterial blooms been vanquished? Phycology (in press).

Kitchell, J. F., L. A. Eby, X. He, D. E. Shindler & R. A. Wright, 1994. Predator-prey dynamics in an ecosystem context. J. Fish Biol. 45 (suppl. A) 209–226.

Lieder, U., 1983a. Die Arten der Untergattung Eubosmina Seligo, 1900 (Crustacea: Cladocera, Bosminidae). Mitt. Zool. Mus. Berlin 59 (2): 195–292.

Lieder, U., 1983b. Revision of the genus Bosmina Baird, 1845 (Crustacea, Cladocera). Int. Rev. Hydrobiol. 68: 121–139.

Lyche, A., B. A. Faafeng & Å. Brabrand, 1990. Predictability and possible mechanisms of plankton response to reduction of planktivorous fish. Hydrobiologia 200/201: 251–262.

McQueen, D. J., M. R. S. Johannes, J. R. Post, T. J. Stewart & D. R. S. Lean, 1989. Bottom-up and top-down impacts on freshwater pelagic community structure. Ecological Monogr. 59: 289–309.

McQueen, D. J., R. France & C. Kraft, 1992. Confounded impacts of planktivorous fish on freshwater biomanipulations. Arch. Hydrobiol. 125: 1–24.

Meijer, M.-L., M. W. de Haan, A. W. Breukelaar & H. Buiteveld, 1990. Is reduction of the benthivorous fish an important cause of high transparancy following biomanipulation in shallow lakes? Hydrobiologia 200/201: 303–315.

Moss, B., 1990. Engineering and biological approaches to the restoration from eutrophication of shallow lakes in which aquatic plant communities are important components. Hydrobiologia 200/201: 367–377.

Persson, A., 1997a. Effects of fish predation and excretion on the configuration of aquatic food webs. Oikos 79: 137–146.

Persson, A., 1997b. Phosphorus release by fish in relation to external and internal load in a eutrophic lake. Limnol. Oceanogr. 42: 577–583.

Persson, L., G. Andersson, S. F. Hamrin & L. Johansson, 1988. Predator regulation and primary production along the productivity gradient of temperate lake ecosystems. In S. R. Carpenter (ed.), Complex Interactions in Lake Communities. Springer-Verlag. New York: 45–65.

Riemann, B., K. Christoffersen, H. J. Jensen, J. P. Müller, C. Lindegaard & S. Bosselmann, 1990. Ecological consequences of a manual reduction of roach and bream in a eutrophic, temperate lake. Hydrobiologia 200/201: 241–250.

Romare, P. & E. Bergman, 1999. Juvenile fish expansion following biomanipulation and the effect on zooplankton. Hydrobiologia 404: 89–97.

Sanni, S. & S. B. Wærvågen, 1990. Oligotrophication as a result of planktivorous fish removal with rotenone in the small, eutrophic, Lake Mosvatn, Norway. Hydrobiologia 200/201: 263–274.

Schriever, P., J. Bøgestrand, E. Jeppesen & M. Søndergaard, 1995. Impact of submersed macrophytes on fish-zooplankton-phytoplankton interactions: large-scale enclosure experiments in a shallow eutrophic lake. Freshwat. Biol. 33: 255–270.

Shapiro, J., V. Lamarra & M. Lynch, 1975. Biomanipulation: an ecosystem approach to lake restoration. In P. L. Brezonik & J. L. Fox (eds), Proceedings of the Symposium on Water Quality Management Through Biological Control. Rep. No. ENV-07-75-1: 85–96.

Shapiro, J., & D. I. Wright, 1984. Lake restoration by biomanipulation: Round Lake, Minnesota – the first two years. Freshwater Biol. 14: 371–393.

Soranno, P. A., S. R. Carpenter & X. He, 1993. Zooplankton biomass and body size. In S. R. Carpenter & J. F. Kitchell (eds), The Trophic Cascade in Lakes. Cambridge University Press, Cambridge: 172–188

Stansfield, J. H., M. R. Perrow, L. D. Tench, A. J. D. Jowitt & A. A. L. Taylor, 1997. Submerged macrophytes as refuges for grazing Cladocera against fish predation: observations on seasonal changes in relation to macrophyte cover and predation pressure. Hydrobiologia 342/343: 229–240.

Strand, J., 1999. The development of submerged macrophytes in Lake Ringsjön after biomanipulation. Hydrobiologia 404: 113–121.

Svensson, J. M., E. Bergman & G. Andersson, 1999. Impact of cyprinid reduction on the benthic invertebrate community and implications for increased nitrogen retention. Hydrobiologia 404: 99–112.

Søndergaard, M., E. Jeppesen, E. Mortensen, E. Dall, P. Kristensen & O. Sortkjær, 1990. Phytoplankton biomass reduction after planktivorous fish reduction in a shallow, eutrophic lake: a combined effect of reduced internal P-loading and increased zooplankton grazing. Hydrobiologia 200/201: 220–240.

Threlkeld, S. T., 1987. Experimental evaluation of trophic-cascade and nutrient-mediated effects of planktivorous fish on plankton community structure. In W. C. Kerfoot & A. Sih (eds), Predation. Direct and Indirect Impacts on Aquatic Communities. University Press of New England, Hanover and London: 172–188.

Hydrobiologia **404:** 89–97, 1999.
L.-A. Hansson & E. Bergman (eds), Nutrient Reduction and Biomanipulation as Tools to Improve Water Quality: The Lake Ringsjön Story.

Juvenile fish expansion following biomanipulation and its effect on zooplankton

Pia Romare & Eva Bergman
Department of Limnology, Insitute of Ecology, Ecology Building, University of Lund, SE-223 62 Lund, Sweden

Key words: 0+ fish abundance, biomanipulation, *Daphnia*, predation pressure

Abstract

In Lake Ringsjön, southern Sweden, we estimated yearly variations in both 0+ and older fish abundances, following the fish reductions in 1988–1992. Between 1987 and 1994, the total planktivore biomass decreased fourfold whereas the abundance of 0+ fish increased twenty times. The resulting change in predation pressure from planktivorous fish on zooplankton increased up to 2–3 times, after the fish reduction. We conclude that after an initial decrease, the predation pressure on zooplankton increased considerably due to the increase in 0+ fish abundance. The expansion in juvenile fish abundance may therefore have been an important factor behind the apparent lack of top-down effect on the zooplankton community following biomanipulation of Lake Ringsjön. The horizontal distribution of 0+ fish in Lake Ringsjön is also discussed.

Introduction

The importance of 0+ (young-of-the-year) fish predation on zooplankton was not considered in early biomanipulation work (Zalewski et al., 1990), and the aim of such activities was to decrease predation pressure on large zooplankton, by reducing the abundance of adult planktivorous fish. In recent years however, the role of 0+ fish has been given more attention (Meijer et al., 1994; Benndorf, 1995; Mehner et al., 1995; Qin & Culver, 1995), and it has been shown that the abundance of 0+ fish generally increases in lakes after biomanipulation (Meijer et al., 1994; Benndorf, 1995). Since most 0+ fish, regardless of species, are planktivorous and may reach high abundances in summer, they are potentially important regulators of the zooplankton community (Cryer et al., 1986; Mills et al., 1987; Whiteside, 1988). It has even been argued that 0+ fish may be responsible for natural dynamics in cladoceran zooplankton abundance during spring and summer (Keast, 1988; Whiteside, 1988; Vijverberg et al., 1990; Zalewski et al., 1990, but see Boersma et al., 1996). Consequently, high abundances of 0+ fish are now considered a major problem in many lakes that have been subject to biomanipulation (Meijer et al., 1994; Benndorf, 1995).

This chapter describes the change in catches and species composition of 0+ fish and older individuals after a fish reduction in Lake Ringsjön, southern Sweden. Annual variation in 0+ abundance was estimated from trawlings carried out in August 1987 and August 1990–1994, while horizontal distribution of 0+ fish and seasonal changes in size were obtained from Bongo trawlings in 1990 and 1991. Stomach analyses were done on fish to determine which size classes consumed zooplankton, and this information was used to model the change in predation pressure on zooplankton after fish reduction. We did this in order to evaluate whether the predation pressure on zooplankton in Lake Ringsjön actually increased after the fish reduction, depending on an increase in 0+ fish abundance.

Study site

Lake Ringsjön is situated in southernmost Sweden (55° 52′ N; 13° 32′ O) and consists of three basins: Sätofta Basin (4.2 km^2, mean depth 3.0 m, maximum depth 17.0 m), Eastern Basin (20.5 km^2, mean depth 6.1 m, maximum depth 16.4 m) and Western Basin (14.8 km^2, mean depth 3.1 m, maximum depth 5.4 m) (Hansson et al., 1999). As is the case with many other lakes in urban and agricultural areas, Lake Ringsjön became eutrophicated during the 1960s and 1970s. In the beginning of the 1980s, a *nutrient reduction programme* was initiated which reduced the external input of nutrients to the lake from phosphorus loadings exceeding 30 t per year to around 10 t per year. About ten years later (1989–1990), a *cyprinid fish reduction programme* (biomanipulation) was carried out in Sätofta Basin removing about 100 t of fish (about 60% of the standing stock). During the winter of 1988, there was a fish-kill in Eastern Basin, eliminating about 500 t of fish (about 80%) (Hamrin, 1999). The biomanipulation in Western Basin, which was performed in 1992, removed about 120 t of fish (about 50%).

Methods

0+ fish were caught by two different methods. Bongo trawling (mesh size 500 μm, net diameter 60 cm, and net length 3 m) was used to study 0+ fish distribution in the surface water and benthic trawling (see Bergman et al., this volume) was used to estimate total fish abundance at the end of the summer.

Bongo trawlings were performed at night in the summer and autumn of 1990 and 1991 in Sätofta Basin and in Eastern Basin. The net was towed behind a small fishing boat at a speed of 1.5–2 m s^{-1} just below the surface. Sampling was performed on four occasions each year; in 1990 in May, June, August and October, and in 1991 in early and late June, August and September. 0+ fish were collected at three locations (nearshore, halfway between the shore and the center of the basin, and at the center of the basin) on all sampling occasions, except those in June 1991 when only one location was sampled. The 0+ fish were immediately preserved in alcohol (70%) and later sorted in the laboratory, where they were identified, counted and measured (distance from tip of snout to end of caudal fin, TL) to the nearest mm. Average numbers caught per sampling date were calculated as the mean of the different locations standardized for 1 min of towing.

Recently, it has been shown that bongo trawling is not an efficient method to catch 0+ fish >20 mm (Thiel, 1996; Wanzenböck et al., 1997; Whiteside & Hatch, 1997) and that bongo catches should not be used to describe seasonal changes in abundance. However, we believe that our bongo data can be used to compare 0+ catches within the same sampling date, assuming avoidance to be of the same magnitude at all locations.

Results from the benthic trawlings in August 1987 and August 1990–1994 were used to estimate relative numbers and biomass (Catch Per Unit Effort, see Bergman et al., 1999) of each species of 0+ fish. When comparing inter-annual changes in the abundance of 0+ fish, catches are expressed as the sum of the catches from 3 m and 5 m tows (catch per 40 min trawling) since 75–100% of all 0+ fish were collected at these depths. When describing the relation between abundance of 0+ fish and older planktivores, and when calculating total predation pressure, we needed to include biomass data from all depths so that depth-integrated catches (catch per 20 min trawling, see Bergman et al., 1999) are used.

Dietary analysis on fish collected in 1987, 1989, 1990 and 1994 were performed. In 1989, fish from the fish reduction trawlings in Sätofta Basin were used. The ingested food organisms were identified to genus or species, and counted. Data on stomach analysis from zander (*Stizostedion lucioperca*), perch (*Perca fluviatilis*), roach (*Rutilus rutilus*) and bream (*Abramis brama*) collected from benthic trawlings was used to establish the extent of planktivory (in%) in each size class. The biomasses of older planktivores (>0+) actually consuming zooplankton, were calculated using biomass data for each size class of fish multiplied by the extent of planktivory. Between 4 and 20 0+ fish per sampling date, from both Bongo and benthic trawlings, were examined to compare the share of *Daphnia* in the stomach zooplankton with the share of *Daphnia* in the zooplankton samples from the lake. For this comparison, we only used 0+ perch and zander since they were the only 0+ species available in early summer trawlings. Zooplankton and water temperature data were collected from routine monthly samplings (Bergman, 1999; Bergman & Bergstrand, 1999).

In order to evaluate whether the increase in 0+ abundance in Lake Ringsjön could have a substantial effect on zooplankton, especially *Daphnia*, we estimated the 'Relative predation pressure' (RPP) from the planktivores in Lake Ringsjön in 1987–1994. The planktivores were divided into 0+ fish and adults (that is, older than 0+). The maximum consumption values (daily ration as % body dry weight) used for 0+ perch (1.0 g body weight) and older planktivores (>100 g body weight) were taken from literature (Lessmark, 1983; Marmulla & Rösch, 1990; Post, 1990; Karjalainen et al., 1997). The maximum consumption by 1.0 g 0+ fish is found to be 4–7 times higher than for the corresponding weight of older planktivores. We estimated RPP in August in Sätofta Basin and Eastern

Basin for all years of the study, assuming that all 0+ fish weighed (wet weight) 1.0 g in August. The RPP values were then estimated by multiplying the values of relative consumption by the biomass of 0+ fish and adult planktivores in August as follows:

$$RPP = (a^*B_A) + (b^*B_{0+}),$$

where a=relative consumption for adults (set to 1), b=relative consumption for 0+ fish that weighed 1.0 g (set to 4 and 7), B_A=biomass of adult fish and B_{0+}=biomass of 0+ fish. As 0+ fish >15–20 mm are able to feed on adult *Daphnia* (Treasurer, 1990; Jachner, 1991; Romare, unpublished data), we included all 0+ in our calculations.

Results

Benthic trawling

The abundance of 0+ fish in August, estimated from benthic trawl catches at 3 m and 5 m depth together, increased gradually following fish reduction in all basins. In 1994, the abundance of 0+ fish was twenty times higher than in 1987 in all basins (Figure 1), and the 0+ fish comprised more than 80% of total fish abundance and more than 20% of the biomass (Figure 2, Bergman et al., 1999).

In Sätofta Basin, high numbers of 0+ fish were caught in 1992 and in 1994 (Figure 1) and 0+ fish made up more than 80% of total fish abundance and 30–50% of the biomass (Figure 2). The high 0+ abundance was due to high numbers of 0+ roach in 1992, and roach and bream in 1994 (Figure 1). In 1987, 1990, 1991 and 1993, 0+ fish made up less than 50% of total fish abundance and less than 6% of the biomass (Figure 2), 0+ catches were then dominated by perch (Figure 1). About 90% of the 0+ fish were found in shallow waters above 5 m during all years.

In Eastern Basin, large numbers of 0+ fish were caught in 1993 and 1994 (Figure 1), they made up more than 80% of the total fish abundance and 10–20% of the biomass (Figure 2). Although the numbers of 0+ fish were much lower in 1991 and 1992, 0+ made up 60% of the total abundance (Figure 2). Perch dominated 0+ catches until 1994 when large numbers of 0+ roach were also caught (Figure 1). Between 75 and 100% of the 0+ fish were found in shallow waters above 5 m depth during all years except in 1991 when most 0+ fish were caught further out.

In Western Basin, large numbers of 0+ fish occurred for the first time in 1994 (Figure 1) and 0+ fish made up 80% of total fish abundance and 20% of the biomass (see Figure 2). As in Eastern Basin, 0+ fish had made up more than 60% of the total fish abundance since 1991, but this was less than 6% of total fish biomass (Figure 2). Perch was the most abundant 0+ fish species until 1992 when the number of roach caught started to increase (Figure 1).

Bongo trawling

The horizontal distribution of 0+ fish differed between Sätofta Basin and Eastern Basin (Table 1). In Sätofta Basin the highest numbers of 0+ fish were always caught away from the shore, whereas in Eastern Basin the highest numbers occurred near-shore (Table 1). These differences were, however, not significant given the high variation in catches between years. It should also be pointed out that 0+ abundance was higher in Sätofta Basin than in Eastern Basin in the surface waters away from the shore (Mann-Whitney U-test, $P<0.05$), while there was no difference in 0+ abundance between the two basins at the near shore locations (Mann-Whitney U-test, NS). Perch was the most abundant 0+ fish in all catches, which corroborates our results from the benthic trawl data from these years. The size of the 0+ fish caught in late June differed from 1990 to 1991, with 0+ perch reaching a mean length of 35 mm in late June 1990, but only about 20 mm in late June 1991 (t-test, $p<0.0001$). By the end of the summer the size difference had almost disappeared, being 50 mm in August 1990 and 42 mm in 1990.

Diet and Relative Predation Pressure (RPP)

Stomach analysis showed that 0+ fish of all species examined were planktivorous during their first summer and that 0+ perch and zander, 30–70 mm in size, ate *Daphnia* throughout the summer in Eastern Basin. The share of *Daphnia* was always higher in the stomach content than in the zooplankton samples, indicating a preference for *Daphnia* (Figure 3). There was no gut analysis done on fish less than 30 mm from Lake Ringsjön, but in a nearby lake *Daphnia* were consumed by roach as small as 17 mm (Romare, unpublished data). Hence, we have assumed that all 0+ fish collected in Lake Ringsjön during this investigation were able to consume *Daphnia*.

The RPP increased during years of high 0+ abundance (Figure 4) in both Sätofta Basin and Eastern Basin. In Sätofta Basin the RPP in August 1991, 1992

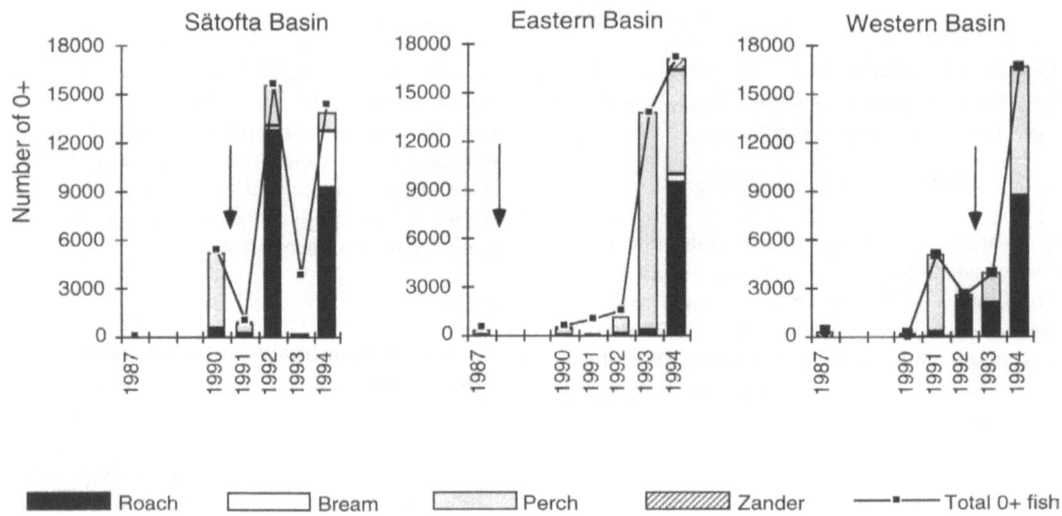

Figure 1. Numbers of 0+ fish collected by benthic trawling in different years in Sätofta Basin (left), Eastern Basin (middle) and Western Basin (right) of Lake Ringsjön. Bars indicate the numbers of all species of 0+ fish caught in the 3 m and 5 m tows summed together, black line indicates total 0+ numbers from all depths towed. Arrows indicate when fish reduction ended.

Table 1. The horizontal distribution of 0+ fish in Sätofta Basin and Eastern Basin expressed as numbers of fish caught during one minute of Bongo trawling, at three different distances from the shore; I=nearshore, II=halfway between the shore and the center of the basin, III=center of the basin. Trawlings were performed in May, June and August 1990, and in August 1991

	Sätofta Basin					Eastern Basin			
	1990			1991		1990			1991
	May	June	Aug.	Aug.		May	June	Aug.	Aug.
I	27	138	21	8	I	1	158	22	33
II	95	685	60	435	II	0	46	6	13
III	405	65	50	767	III	3	38	1	1

and 1994 was 2–3 times higher than in the years before fish reduction. The RPP in Eastern Basin first decreased following the fish reduction in 1988 to its lowest value in 1991, and then increased again (Figure 4). The RPP in August 1994 in Eastern Basin, was however, still lower than in the year before fish reduction. When high RPP values (here defined as higher than 75) occurred after fish reduction, more than 50% of the RPP was due to the consumption of 0+ fish except in Sätofta Basin in 1991 (Figure 4).

Discussion

Benthic and Bongo trawling

An increase in 0+ fish, after fish reduction was found in shallow lakes in Denmark and the Netherlands, where 0+ numbers began to increase 1–3 years after the fish reduction (Meijer et al., 1994). In Lake Ringsjön, the response time for 0+ fish differed between basins and seemed to depend on degree of fish reduction. In Sätofta Basin and in Western Basin, the large increase in 0+ numbers occurred two years after a 50–60% fish reduction, whereas in Eastern Basin, the increase occurred six years after an 80% fish reduction. Some authors have argued that biomanipulation by stocking and protecting piscivorous fish may result in

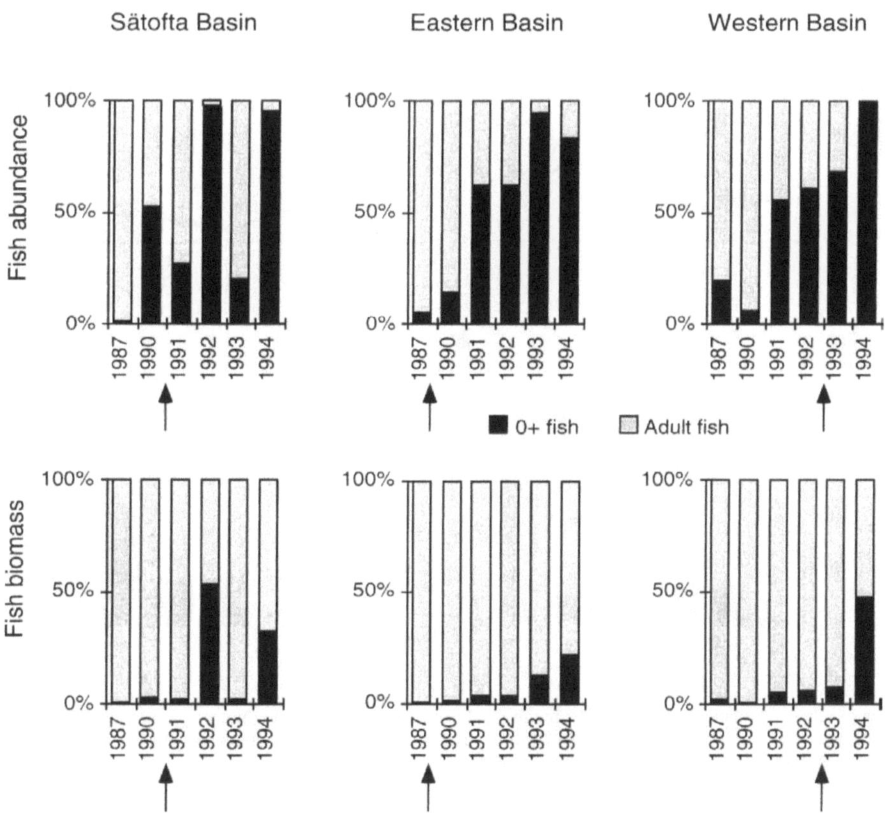

Figure 2. Proportion of juvenile and adult fish in the benthic trawl catches in different years in Sätofta Basin (left), Eastern Basin (middle) and Western Basin (right) of Lake Ringsjön, calculated for both fish abundance (top) and fish biomass (bottom). Arrows indicate when fish reduction ended.

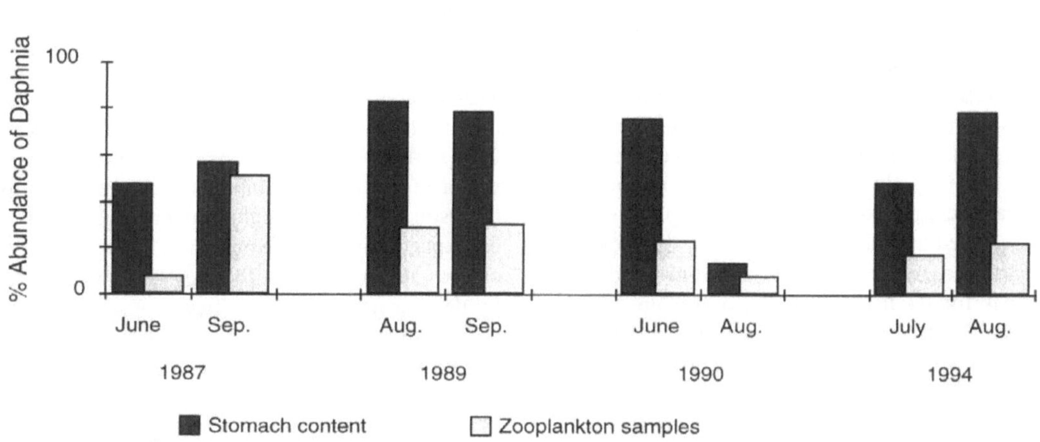

Figure 3. Proportion of *Daphnia* numbers in stomachs of 0+ fish and in zooplankton samples on two occasions each year in 1987, 1989, 1990 and 1994.

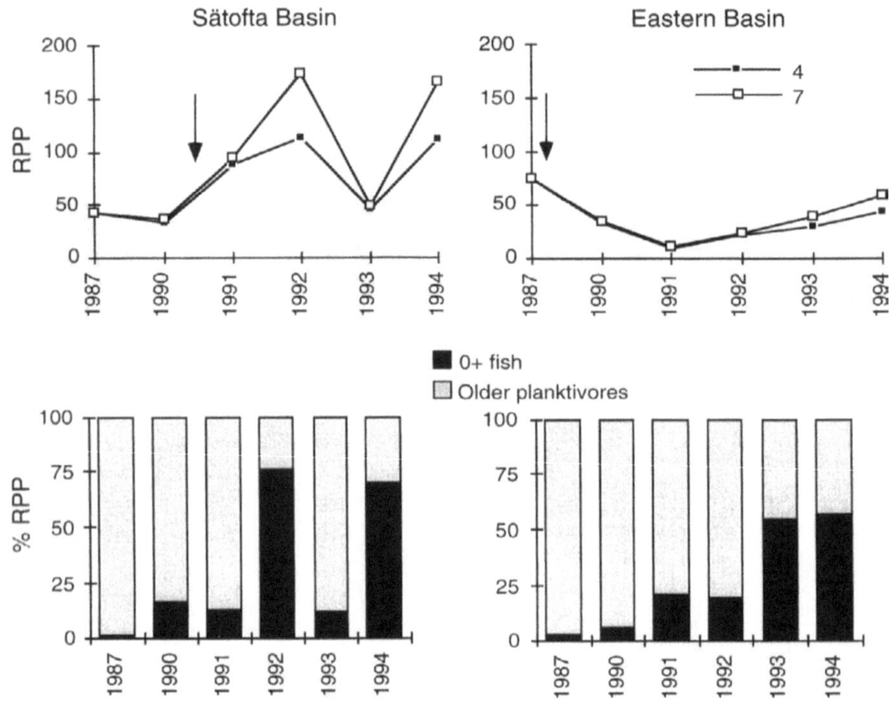

Figure 4. Estimates of relative predation pressure (RPP) by all planktivores (above) – the maximum consumption for 0+ fish being 4 and 7 times higher than for older fish – and proportion of RPP (%) of 0+ fish in relation to older planktivores (below) in different years in Sätofta Basin (left) and Eastern Basin (right). Arrows indicate when fish reduction ended.

higher offspring numbers of piscivorous fish and hence high numbers of planktivorous 0+ fish (Qin & Culver, 1995; Mehner et al., 1995; Mehner et al., 1997). In Lake Ringsjön, the biomanipulation did not include any stocking of piscivorous fish but only the reduction of adult planktivorous fish (see Hamrin, 1999), this fish reduction still lead to a compensatory increase in both perch and roach 0+ numbers. Abundance of 0+ fish is influenced by lake morphometry, macrophyte cover and the presence of piscivores (Carpenter & Lodge, 1986; Lodge et al., 1988). The amount of 0+ fish may increase with increased macrophyte cover due to increased availability of spawning sites for the adults and increased availability of refuges and foraging areas for 0+ fish (Carpenter & Lodge, 1986; Lodge et al., 1988). Because there was only a marginal increase in macrophyte cover after the fish reduction (Strand, 1999), the probable mechanism behind the increase in 0+ numbers, is the decreased competition from adult roach making food resources (zooplankton) available for the 0+ fish. It has been shown that decreasing competition from older roach affects 0+ perch biomass positively (Persson & Greenberg, 1990). At the same time piscivore numbers may have been too low to keep 0+ numbers down during summer.

Species composition, as reflected in the benthic catches at 3 m and 5 m, differed among basins. Hence, we mainly caught 0+ roach in Sätofta Basin and 0+ perch in Eastern Basin. The reproductive potential for planktivores is much higher than for piscivores in productive lakes (Persson et al., 1991; Post & Kitchell, 1997) and 0+ roach is considered competitively superior in unstructured habitats compared to 0+ perch (Persson, 1991), thus high numbers of 0+ roach could be expected in Lake Ringsjön. The presence of piscivores however, may affect perch biomass positively by decreasing competition from 0+ roach when they are both forced into vegetated habitats (Persson, 1991, 1997). This may have been the case in Eastern Basin where the proportion of piscivores was higher than in Sätofta Basin (Bergman et al., 1999).

Distribution and migration of 0+ fish is also influenced by lake morphometry, macrophyte cover and the presence of piscivores (Carpenter & Lodge, 1986;

Lodge et al., 1988; Diehl & Eklöv, 1995). Results from Bongo trawlings indicate that the total number of 0+ fish was higher in the surface waters away from shore in Sätofta Basin, than in Eastern Basin and this may relate to macrophyte beds extending further away from shore in the shallower Sätofta Basin (Strand, 1999). Since Bongo trawling was performed at night, the higher abundance of 0+ fish in Sätofta Basin might also reflect night time migrations of 0+ out of the vegetation, as reported by Jachner (1991). Such diel migrations are more likely to take place in Sätofta Basin than in Eastern Basin, both because of the shorter distance between the open water and macrophyte beds (Lodge et al., 1988), and because of the higher proportion of piscivores in Eastern Basin, which may force the 0+ fish to stay in the littoral zone (Diehl & Eklöv, 1995).

The lower early summer temperature in 1991 rather than in 1990, could explain the smaller size of 0+ in 1991. Mean surface temperature in May and June, as recorded at the routine sampling occations, was 3–4 °C lower in 1991, than in the years 1988–1990 and 1992–1993. Interannual variation in size of 0+ fish has been shown to be significantly related to interannual variation in temperature (Mooij et al., 1994). Several studies have shown that years of low early summer temperatures coincide with years of small-sized 0+ fish in early summer. This has been seen in Lake Finjasjön (Hamrin, unpublished data) and in Lake Vombsjön (Romare, unpublished data) in southern Sweden , as well as in Lake Constance in Germany (Wang & Eckmann, 1994) and Lake Tjeukemeer in the Netherlands (Spaak & Hoekstra, 1994).

Did RPP-values increase after the fish reduction?

The expected top-down effect of lower planktivore biomass leading to increased zooplankton grazing on phytoplankton may fail to occur as a result of an increase of 0+ fish, after a fish reduction (Benndorf, 1995). In some biomanipulated shallow lakes in Denmark and The Netherlands, such increases of 0+ abundance have only been temporary (Meijer et al., 1994), whereas in other lakes the problem has persisted and some have began a return to their pre-biomanipulated state (Meijer et al., 1994; Benndorf, 1995). In Lake Ringsjön, the increased abundance of 0+ fish, after the fish reduction, resulted in an increase in predation pressure on the zooplankton community based on RPP values, which may explain the lack of a top-down effect on the zooplankton community (Bergman & Bergstrand, 1999).

The model we used to estimate RPP from 0+ fish and older planktivores is based on two variables. The first variable is the abundance of 0+ fish, which was derived from the benthic trawlings conducted in August each year. Although we lack data on absolute numbers of 0+, we used the same method each year and thus have data allowing us to make an interannual comparison for each basin. The second variable is maximum daily ration or maximum daily consumption, and these values are derived from numerous experiments in the laboratory and in the field (Lessmark, 1983; Marmulla & Rösch, 1990; Post, 1990), all showing similar results. We believe that RPP in August may be modeled in this simple way since all 0+ fish have reached the size – in Lake Ringsjön 50–60 mm – at which they may ingest all sizes of zooplankton prey (Van Densen, 1985; Treasurer, 1990; Jachner, 1991; Taleb et al., 1994; Qin & Culver, 1995) and maximum daily ration does not change much in that size class (Marmulla & Rösch, 1990). In Sätofta Basin, RPP increased twice following the fish reduction, when the lower consumption value was used and three times when the higher consumption value was used (Figure 4). Even though the RPP in Eastern Basin has not returned to the value of 1987, RPP did increase 4–5 times between 1991 and 1994. In 1991, the increase in RPP in Sätofta Basin was mainly due to high planktivore biomass, probably because planktivore fish migrated into Sätofta Basin from Eastern Basin (Bergman et al., 1999). In 1992–1994 however, total planktivore biomass was much lower than before the fish reduction in both Sätofta Basin and Eastern Basin, and still RPP values increased. The RPP model thus shows that when the portion of 0+ fish biomass increases, total predation pressure on *Daphnia* may increase due to the higher relative consumption of 0+ fish. Since *Daphnia* are considered a key species among zooplankters in controlling phytoplankton biomass (Kerfoot, 1987), it may be necessary to minimize the impact of 0+ fish on *Daphnia* in order to achieve a clear lake after biomanipulation.

Acknowledgement

This work was funded by the Swedish National Board of Fisheries, AB Sydvatten and by the Swedish National Environmental Protection Board. It was also funded by the local authorities in the districts of Eslöv, Hörby and Höör, and by the county administrative board in Malmöhus county. We thank L.-A. Hansson and L. Greenberg for helpful comments on previous drafts of this paper, and the reviewers I. J. Winfield and T. Mehner for valuable comments.

References

Benndorf, J., 1995. Possibilities and limits for controlling eutrophication by biomanipulation. Int. Rev. ges. Hydrobiol. 80: 519–534.

Bergman, E. & E. Bergstrand, 1999. Lack of a top-down effect on the zooplankton community after a cyprinid fish reduction. Hydrobiologia 404: 77–87.

Bergman, E., 1999. Changes in the nutrient load and lake water chemistry in Lake Ringsjön, southern Sweden, from 1966 to 1996. Hydrobiologia 404: 9–18.

Bergman, E., S. F. Hamrin & P. Romare, 1999. The effects of cyprinid reduction on the fish community. Hydrobiologia 404: 65–75.

Boersma, M., O. F. R. van Tongeren & W. M. Mooij, 1996. Seasonal patterns in the mortality of *Daphnia* species in a shallow lake. Can. J. Fish. aquat. Sci. 53: 18–28.

Carpenter, S. R. & D. M. Lodge, 1986. Effects of submersed macrophytes on ecosystem processes. Aquat. Bot. 26: 341–370.

Cryer, M., G. Peirson & C. R. Townsend, 1986. Reciprocal interactions between roach, *Rutilus rutilus*, and zooplankton in a small lake: prey dynamics and fish growth and recruitment. Limnol. Oceanogr. 31: 1022–1038.

Diehl, S. & P. Eklöv, 1995. Effects of piscivore-mediated habitat use on resources, diet, and growth of perch. Ecology 76: 1712–1726.

Hamrin, S. F., 1999. Planning and execution of the fish reduction in Lake Ringsjön. Hydrobiologia 404: 59–63.

Hansson, L.-A., M. Enell & E. Bergman, 1999. Lake Ringsjön: its catchment area, its history and its importance. Hydrobiologia 404: 1–7.

Jachner, A., 1991. Food and habitat partitioning among juveniles of three fish species in the pelagial of a mesotrophic lake. Hydrobiologia 226: 81–90.

Karjalainen, J., D. Miserque & H. Huuskonen, 1997. The estimation of food consumption in larval and juvenile fish: experimental comparison of two methods. J. Fish Biol. (Suppl.) in press.

Keast, K., 1988. Planktivory in larval, juvenile, and adult planktivores: resource division in a small lake. Verh. Int. Ver. Limnol. 23: 1692–1697.

Kerfoot, W. C., 1987. Cascading effects and indirect pathways. In W. C. Kerfoot & A. C. Sih (eds), Predation: Direct and Indirect Impacts on Aquatic Communities. The University Press of New England, Hanover (N.H.): 57–70.

Lessmark, O., 1983. Competition between perch (*Perca fluviatilis*) and roach (*Rutilus rutilus*) in south Swedish lakes. Ph. D. thesis, Univ. of Lund, Lund.

Lodge, D. M., J. W. Barko, D. Strayer, J. M. Melack, G. G. Mittelbach, R. W. Howarth, B. Menge & J. E. Titus, 1988. Spatial heterogeneity and habitat interactions in lake communities. In S. R. Carpenter (ed), Complex Interactions in Lake Communities. Springer-Verlag, New York: 181–208.

Marmulla, G. & R. Rösch, 1990. Maximum daily ration of juvenile fish fed living natural zooplankton. J. Fish. Biol. 36: 789–801.

Mehner, T., H. Schultz & R. Herbst, 1995. Interactions of zooplankton dynamics and diet of 0+ perch (*Perca fluviatilis* L.) in the top-down manipulated Bautzen Reservoir (Saxony, Germany) during summer. Limnologica 25: 1–9.

Mehner, T., M. Plewa, S. Hülsmann, H. Voigt & J. Benndorf, 1997. Age-0 fish predation on daphnids – spatial and temporal variability in the top-down manipulated Bautzen Reservoir, Germany. Arch. Hydrobiol. Spec. Issues Advanc. Limnol. 49: 13–25.

Meijer, M-L., E. Jeppesen, E. van Donk, B. Moss, M. Scheffer, E. Lammens, E. van Nes, J. A. van Berkum, G. J. De Jong, B. A. Faafeng & J. P. Jensen, 1994. Long-term responses to fish-stock reduction in small shallow lakes: interpretation of five-year results of four biomanipulation cases in The Netherlands and Denmark. Hydrobiologia 275/276: 457–466.

Mills, E. L., J. L. Forney & K. J. Wagner, 1987. Fish predation and its cascading effect on the Oneida lake food chain. In W. C. Kerfoot & A. C. Sih (eds), Predation: Direct and Indirect Impacts on Aquatic Communities. The University Press of New England, Hanover (N.H.): 118–131.

Mooij, W. M., E. H. R. R. Lammens & W. L. T. van Densen, 1994. Growth rate of 0+ fish in relation to temperature, body size, and food in shallow eutrophic lake Tjeukemeer. Can. J. Fish. Aquat. Sci. 51: 516–526.

Persson, L. & L. A. Greenberg, 1990. Juvenile competitive bottlenecks: the perch (*Perca fluviatilis*) – roach (*Rutilus rutilus*) interaction. Ecology 71: 44–56.

Persson, L., 1991. Behavioral response to predators reverses the outcome of competition between prey species. Behav. Ecol. Sociobiol. 28: 101–105.

Persson, L., 1997. Competition, predation and environmental factors as structuring forces in freshwater fish communities: Sumar (1971) revisited. Can. J. Fish Aquat. Sci. 54: 85–88.

Persson, L., S. Diehl, L. Johansson, G. Andersson & S. F. Hamrin, 1991. Shifts in fish communities along a productivity gradient of temperate lakes – patterns and the importance of size-structured interactions. J. Fish. Bio. 38: 281–293.

Post, D. M. & J. F. Kitchell, 1997. Trophic ontogeny and life history effects on interactions between age-0 fishes and zooplankton. Arch. Hydrobiol. Spec. Issues Advanc. Limnol. 49: 1–12.

Post, J. R., 1990. Metabolic allometry of larval and juvenile yellow perch (*Perca flavescens*): in situ estimates and bioenergetic models. Can. J. Fish. Aquat. Sci. 47: 554–560.

Qin, J. & D. A. Culver, 1995. Effect of young-of-the-year walley (Percidae: *Stizostedion vitreum*) on plankton dynamics and water quality in ponds. Hydrobiologia 297: 217–227.

Spaak, P. & J. R. Hoekstra, 1994. Population dynamics of a *Daphnia* hybrid species complex in Tjeukemeer. In P. Spaak, Genetical Ecology of a coexisting *Daphnia* hybrid species complex. Ph. D. thesis, University of Utrecht, the Netherlands: 97–116.

Strand, J., 1999. The development of submerged macrophytes in Lake Ringsjön after biomanipulation. Hydrobiologia 404: 113–121.

Taleb, H., P. Reyes-Marchant & N. Lair, 1994. Effect of vertebrate predation on the spatio-temperal distribution of cladocerans in a temperate eutrophic lake. Hydrobiologia 294: 117–128.

Thiel, R., 1996. The impact of fish predation on the zooplankton community in a southern Baltic bay. Limnologica 26: 123–137.

Treasurer, J. W., 1990. The food and daily consumption of lacustrine 0+perch, *Perca fluviatilis* L. Freshwat. Biol. 24: 361–374.

van Densen, W. L. T., 1985. Feeding behaviour of major 0+fish species in a shallow, eutrophic lake (Tjeukemeer, The Netherlands). Z. Angew. Ichthyol. 2: 49–70.

Wang, N. & R. Eckmann, 1994. Distribution of perch (*Perca fluviatilis* L.) during their first year of life in Lake Constance. Hydrobiologia 277: 135–143.

Wanzenböck, J., J. Matena & J. Kubecka, 1997. Comparison of two methods to quantify pelagic early life stages of fish. Arch. Hydrobiol. Spec. Issues Advanc. Limnol. 49: 117–124.

Whiteside, M. C. & J. T. Hatch, 1997. Quantitative sampling techniques for age-0 fish from diverse lake habitats. Arch. Hydrobiol. Spec. Issues Advanc. Limnol. 49: 99–116.

Whiteside, M. C., 1988. 0+ fish as major factors affecting abundance patterns of littoral zooplankton. Verh. Int. Ver. Limnol. 23: 1710–1714.

Vijverberg, J., M. Boersma, W. van Densen, W. Hoogenboezem, E. H. R. R. Lammens & W. M. Mooij, 1990. Seasonal variation in the interactions between piscivorous fish, planktivorous fish and zooplankton in a shallow eutrophic lake. Hydrobiologia 207: 279–286.

Zalewski, M., B. Brewinska-Zaras, P. Frankiewicz & S. Kalinowski, 1990. The potential for biomanipulation using fry communities in a lowland reservoir: concordance between water quality and optimal recruitment. Hydrobiologia 200/201: 549–556.

Hydrobiologia **404:** 99–112, 1999.
L.-A. Hansson & E. Bergman (eds), Nutrient Reduction and Biomanipulation as Tools to Improve Water Quality: The Lake Ringsjön Story.

Impact of cyprinid reduction on the benthic macroinvertebrate community and implications for increased nitrogen retention

Jonas M. Svensson[1], Eva Bergman[1] & Gunnar Andersson[2]
[1]*Department of Limnology, Insitute of Ecology, Ecology Building, University of Lund, SE-223 62 Lund, Sweden*
[2]*County Administration of Scania, SE-205 15 Malmö, Sweden*

Key words: benthic macroinvertebrates, fish reduction, biomanipulation, bioturbation, nutrient retention

Abstract

The benthic macroinvertebrate community in a severely eutrophicated South Swedish lake, Lake Ringsjön, was found to re-establish after the lake was subjected to a substantial reduction of cyprinids (mainly bream and roach). Being totally dominated by chironomids and oligochaetes during the 1980s, the lake became repopulated by groups like Amphipoda, Ephemeroptera, Coleoptera and Mollusca after the fish reduction in 1988–1992. These groups appeared in numbers similar to that found in 1969, before the lake entered its worst stage of eutrophication. The re-establishment of the invertebrates mainly occurred in the littoral parts of the lake. The profundal community was still dominated by chironomids and oligochaetes after the fish reduction. The increase in numbers of benthic macroinvertebrates in the littoral zone of the lake was likely an effect of lowered predation pressure from fish. The potential role of an increased macroinvertebrate abundance due to fish removal and its effects on recirculation of nutrients over the sediment–water interface is discussed. It was observed that average total nitrogen retention in Lake Ringsjön before and after the reduction was 59±5% and 82±15%, respectively.

Introduction

Reduction of cyprinid fish populations in eutrophic lake ecosystems has become an important tool for improving lake water quality, such as reducing the frequency of algae by reducing the predation pressure from fish on zooplankton (Gophen, 1990; Shapiro, 1990). Effects of large scale reductions of cyprinid fish in eutrophic lake ecosystems on water quality and zoo- and phyto-plankton communities have been thoroughly described in several papers (cf. Andersson, 1984; Jeppesen et al., 1990a,b; Riemann et al., 1990; Hamrin, 1993). However, more sparse and seldom reported are the effects of a managed fish community on benthic macroinvertebrate abundance and species richness. Likewise, benthic macroinvertebrates are generally excluded when modelling trophic-level interactions in lakes (Carpenter & Kitchell, 1988; Persson et al., 1988; Gophen, 1990; Hosper & Jagtman, 1990; Shapiro, 1990).

Generally, an increase in nutrient supply will cause an increase in the abundance of chironomids and oligochaetes, whereas ephemeropterans will decrease in abundance (Jonasson, 1969). It has also been reported that large populations of fish can reduce the abundance of benthic animals like snails, clams, crustaceans and chironomids (Svärdson, 1976; Persson 1983a,b; Andersson, 1984; Lammens et al., 1985; Kajak, 1988; Walls et al., 1990). Although possibly dubious to extrapolate to large lakes, field enclosure experiments have shown that benthic macroinvertebrate predators are affected by fish predation, while their main prey, chironomids, may increase in abundance as a result of decreased predation (Persson & Greenberg, 1990; Diehl, 1992; Bergman & Greenberg, 1994). Simultaneously, and as a direct consequence of fish consuming invertebrates, birdlife may be impoverished (Nilsson, 1972; Andersson, 1981).

Benthic macroinvertebrates are not only an important food resource for fish and waterfowl (cf. Nilsson, 1972) but may also enhance circulation of inorganic phosphorus and nitrogen species over the sediment-water interface and increase bacterial activity in the sediment (Boström et al., 1982; Fukuhara & Sakamoto

1987, 1988; Andersson et al., 1988; Aller, 1988; Kristensen, 1988).

The water quality of Lake Ringsjön, situated in southern Sweden, has detoriated dramatically during this century. In the later part of the 1960s, the situation became intolerable with a halved Secchi depth and quadrupled concentrations of total phosphorus relative to the situation before the 1960s. In 1977–1980, the external phosphorus load reached 0.75 g P m^{-2} yr^{-1}. Soon after, all municipal water was treated according to new environmental standards (Hamrin, 1993) and measures were taken among the cattle farmers living adjacent to the lake, to lower nutrient leakage. Despite the fact that these measures lowered external phosphorus inputs, Secchi depth did not improve. The decision was therefore taken to perform an extensive reduction of the cyprinid population of the lake. From the spring of 1988 to the autumn of 1992, the three basins of Lake Ringsjön were subject to a substantial reduction of cyprinids. Since benthic macroinvertebrates is a food resource for fish and waterfowl and may also act as a potential link for increased nutrient circulation, we studied changes in the invertebrate population in relation to the fish manipulation. Data on benthic macroinvertebrates were avaliable for various years between 1943 and 1996 and we attempt to relate these data to the different states Lake Ringsjön experienced during this time. We were particularly interested in the invertebrate response to the fish manipulation and in the coupling between change in invertebrate abundance and nutrient removal.

Material and methods

Study site

Lake Ringsjön is situated in southernmost Sweden (N55° 52′; O13° 32′) and consists of three basins: Sätofta Basin (4.2 km^2, mean depth 3.0 m, maximum depth 17.0 m), Eastern Basin (20.5 km^2, mean depth 6.1 m, maximum depth 16.4 m) and Western Basin (14.8 km^2, mean depth 3.1 m, maximum depth 5.4 m) (Hansson et al., 1999). As many other lakes in urban and agricultural areas, Lake Ringsjön became eutrophicated during the 1960s and 1970s. In the beginning of the 1980s, a *nutrient reduction programme* was initiated which reduced the external input of nutrients to the lake from phosphorus loadings exceeding 30 tons per year to around 10 tons per year. About ten years later (1989–1990), a *cyprinid fish reduction programme* (biomanipulation) was carried out in Sätofta Basin removing about 100 tons of fish (about 60% of the standing stock). During the winter of 1988, there was a fish-kill in Eastern Basin, eliminating about 500 tons of fish (about 80%) (Hamrin, 1999). The biomanipulation in Western Basin, which was performed in 1992, removed about 120 tons of fish (about 50%).

Between 1943 and 1996, samples of macroinvertebrates in Lake Ringsjön were taken on eight occasions during four different investigations (Figure 1). The scopes of these investigations differed and consequently the sampling methods have not been the same. Below, we describe the sampling methods used and the data available to us from the different studies.

The purpose of the first study was to relate the macroinvertebrate fauna to different types of sediment of South Swedish lakes (Brundin, 1949). Invertebrates were collected in Western Basin from vegetation-free bottoms at 4.3–4.5 m depth, and from *Ceratophyllum*-covered sediments at 3.5 m. The samples were taken with an Ekman grab and sieved through a 0.6 mm net. The invertebrates were sorted live and were then preserved in ethanol. Abundance of the different species is reported. The purpose of the second study, performed in 1969 (Nilsson, 1972), was to compare the distribution of diving waterfowl with the distribution of food and to compare food intake of waterfowl with the composition of the invertebrate fauna and the amount of food avaliable. Benthic invertebrate fauna was collected at 89 stations in Western Basin. Between 3 and 10 replicates (Lillö bay; see Figure 1) were taken at each station with an Ekman grab, and the sediment was passed through a 1 mm sieve. Benthic invertebrates retained by the sieve were preserved in 80% alcohol and sorted at the laboratory (Nilsson, 1972). Data on abundance and wet weight for different taxonomic groups are reported in Nilsson (1972).

The third study, performed in 1982 (Andersson, G., unpubl.), was a direct comparison with the study by Nilsson (1972) to investigate the macroinvertebrate response to eutrophication. The benthic invertebrate fauna was sampled at 9 stations in Lillö Bay in Western Basin, according to Nilsson (1972; see Figure 1). At each station 5 samples were taken with an Ekman grab, the sediment was passed through a 1 mm sieve and the benthos left on the sieve was preserved in 80% ethanol and sorted in the laboratory. Data on abundance and wet weight for the different taxonomic groups are reported here.

The latest study investigated the macroinvertebrate community before and after the cyprinid reduction in Lake Ringsjön. Invertebrates were collected with an

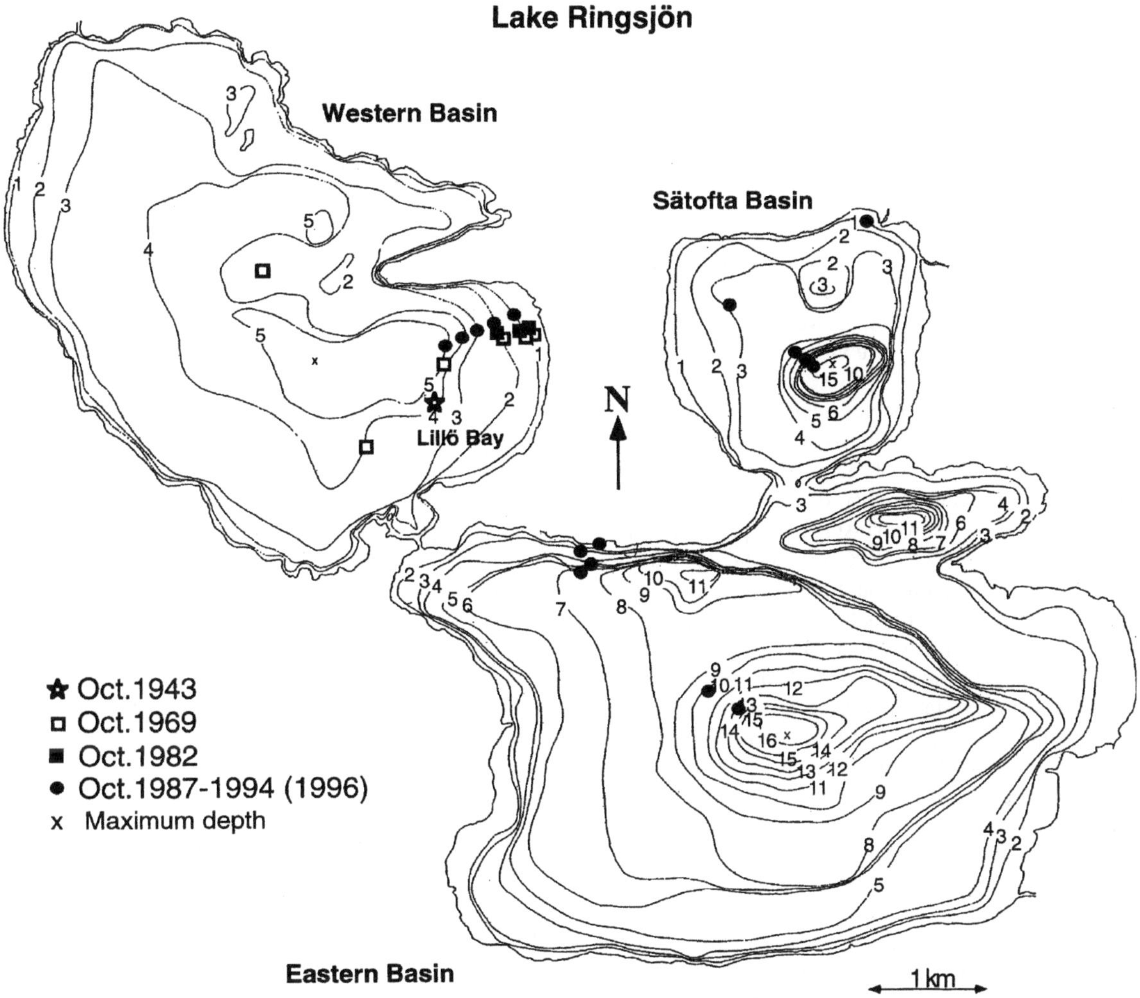

Figure 1. Bathymetric map of Lake Ringsjön. Depth isopleths and sampling stations for benthic invertebrate fauna are indicated in the different basins; Sätofta, Eastern Basin and Western Basin, between the years 1943 and 1996.

Ekman grab from depths of 1, 2, 3, 4 and 5 m in Western Basin and from depths of 1, 3, 5, 7, 10 and 13 m from Sätofta Basin and Eastern Basin in 1987–1996. In 1996, however, benthic invertebrate fauna was only collected in Western Basin. 4–5 samples were taken at each depth, the sediment was passed through a 0.55 mm sieve and the benthos retained on the sieve were transferred to plastic containers and kept cool until sorting. The fauna was sorted live into taxonomic groups and preserved in 80% ethanol. In the laboratory, the samples were dried, burned and weighted. Data on abundance and ash-free dry weight for the different taxonomic groups were reported. In 1996, dry weight was also reported. The diet of perch was analysed for data collected in 1987, 1989, 1990 and 1994, the diet of roach was based on 1987, 1989 and 1994 collections, and the diet of bream was based on 1987 and 1994 collections. In 1987, fish for stomach analyses were collected from all basins. In 1989, fish from Sätofta Basin were used, and in 1990 and 1994 fish from Eastern Basin were used. All size classes except young-of-the-year were analyzed. The number of the different prey items in the stomachs was counted, and the proportion of stomachs with chironomids, ceratopogonids or other taxa was calculated. Empty stomachs were omitted from analyses. A rough estimate of the predation pressure was performed based on biomass of macroinvertebrates found in the stomachs of perch, roach and bream and the biomass of the fish species in the various years.

Statistic comparisons

Since the different investigators used sieves with different mesh-sizes we have not compared the whole material statistically. Below, we describe how the different investigations were viewed and treated.

The sampling from 1943 (Brundin, 1949), was used as a reference for species richness from the period when the lake was a productive, clearwater lake. The samplings from 1969 to 1982 represent the situation just before the water quality deteriorated and when the lake was at its worst stage, respectively. The samplings from 1987 to 1996 represent the period before and after fish reduction. Samplings in 1987 were performed before fish reduction and samplings 1988–1996 occurred after fish reduction in Sätofta Basin and Eastern Basin. In Sätofta Basin, data from 1990 was excluded from the statistics since it was the year of the fish manipulation. In Western Basin, 1987–1990 was before, and 1994–1996 after. Data from 1992 has been excluded since it is the year of the fish manipulation in this basin.

To estimate the benthic fauna composition in the entire lake, depth-weighted means were calculated using abundance data and the weighting factors listed in Table 1. The weighting factors represent the respective proportion of each depth of the total bottom area (total bottom area = 1). For a comparison of the years 1987–1996, data were analyzed using the *t*-test and ANOVA. The data from 1943, 1969 and 1982 was not included in any statistical analyses.

Table 1. Lake area factors used to calculate depth-weighed means for Western, Eastern and Sätofta Basins of Lake Ringsjön

Sampling site	Representative depth (m)	Lake area factor
Western Basin		
1 m	0–1	0.071
2 m	1–2	0.089
3 m	2–3	0.182
4 m	3–4	0.420
5 m	beyond 4	0.238
Eastern Basin		
1 m	0–2	0.113
3 m	2–4	0.181
5 m	4–6	0.222
7 m	6–8	0.188
10 m	8-11	0.222
13 m	beyond 11	0.074
Sätofta Basin		
1 m	0–2	0.272
3 m	2–4	0.546
5 m	4–6	0.097
7 m	6–8	0.025
10 m	8–11	0.028
13 m	beyond 11	0.031

Results

Number of faunal groups

In Western Basin, the total number of faunal groups first decreased and then increased over the period 1943–1996 (Figure 2). Although we lack data from 1943 to 1982 in Sätofta Basin and from 1943–1969 in Eastern Basin, the same pattern was observed in these basins for the latter part of the period (Figure 2). The sampling from 1943 is exclusively from a depth of 4.3–4.5 m, but we assume that the benthic fauna found at this depth also occurred in the shallower parts of Western Basin.

The total number of macroinvertebrate groups found in 1943 and 1969 were similar (Figure 2), but the diversity in the deeper areas in Western Basin decreased markedly between these years. Predation-sensitive groups like amphipods, ephemeropterans, coleopterans and molluscs were eliminated from deeper areas and displaced to shallow areas of Western Basin. These groups were totally absent from Western Basin in 1982, 1987 and 1990, and did not occur at any sampling station. Some re-establishment was observed in the littoral zone in 1992, and in 1994 and 1996 the predation-sensitive groups seemed to return to levels similar to those found in 1969. In the profundal zone, chironomids and oligochaetes dominated during all years they were measured.

A two-way ANOVA was performed for four faunal groups, chironomids, oligochaetes, ceratopogonids and other taxa (see below). In all cases, the ANOVA showed an effect of depth, year (contrasted before and after the fish reduction) and also depth∗year in all three basins separatly ($p < 0.01$ in all cases). Because

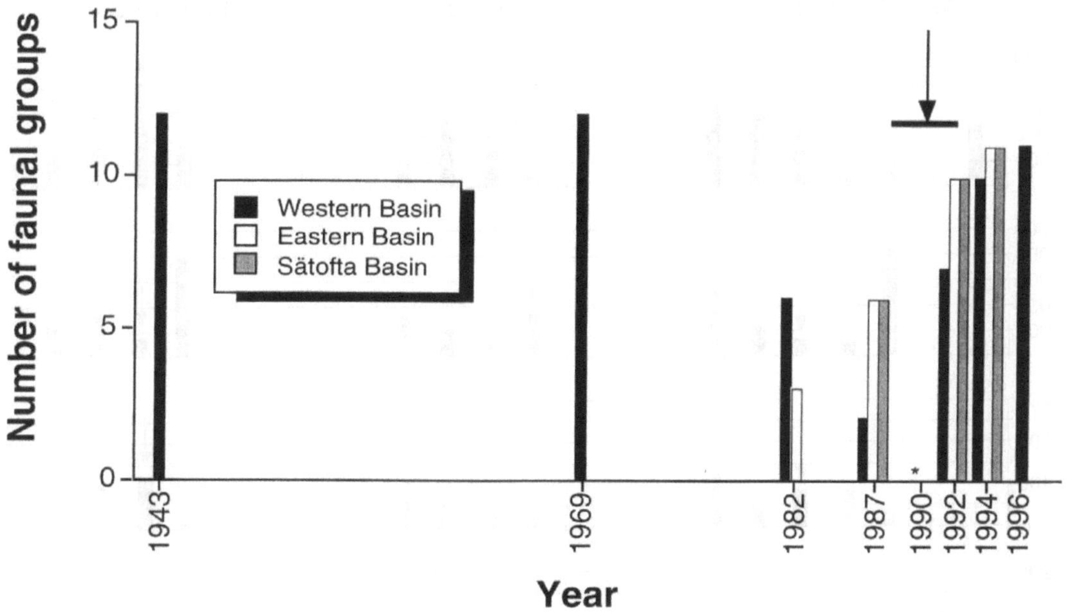

Figure 2. Number of faunal groups represented in the Lake Ringsjön basins between the years 1943 and 1996. Note that sampling was only performed from 1987 to 1994 in Sätofta and Eastern Basins, and that data from 1990 is not available from any basin. Arrow indicates period of fish reduction.

of the significant effect of the interaction term we performed separate analyses for year and depth on all four groups.

Chironomids

Total chironomid abundance (ln-transformed), based on depth-weighted values for 1987–1996, increased significantly after the fish reduction in both Eastern Basin and Western Basin, but not in Sätofta Basin (Figure 3, Table 2). Separate analyses for each depth revealed an increase at 1 m depth after the fish reduction in Sätofta Basin (t=3.2, p<0.01, df=14; Figure 4). In Eastern Basin, chironomid abundance increased only at a depth of 5 m after fish reduction (t=2.94, p<0.01, df=19). In Western Basin, chironomid abundance increased at every depth (t=2.17 to 4.71, p<0.001 to <0.05, df=17), except at 4 m.

Oligochaetes

Oligochaete abundance varied markedly between years (Figure 3). Depth-weighted values of abundance (ln-transformed data 1987–1996) showed that oligochaetes decreased in Sätofta Basin and Eastern Basin

Table 2. Differences between abundances of some selected invertebrate taxa before and after the fish reduction (Student's t-test). Data based on depth-weighed means. NS=differences not significant at the 5% level

	df	*t*	*p*
Western Basin			
Chironomids	16	6.467	<0.001
Oligochaetes	16	1.087	NS
Ceratopogonids	16	1.311	NS
Other groups	16	6.667	<0.001
Eastern Basin			
Chironomids	18	3.262	<0.010
Oligochaetes	18	4.215	<0.001
Ceratopogonids	18	3.007	<0.010
Other groups	18	4.618	<0.001
Sätofta Basin			
Chironomids	13	1.115	NS
Oligochaetes	13	5.800	<0.001
Ceratopogonids	13	2.568	<0.050
Other groups	12	5.049	<0.001

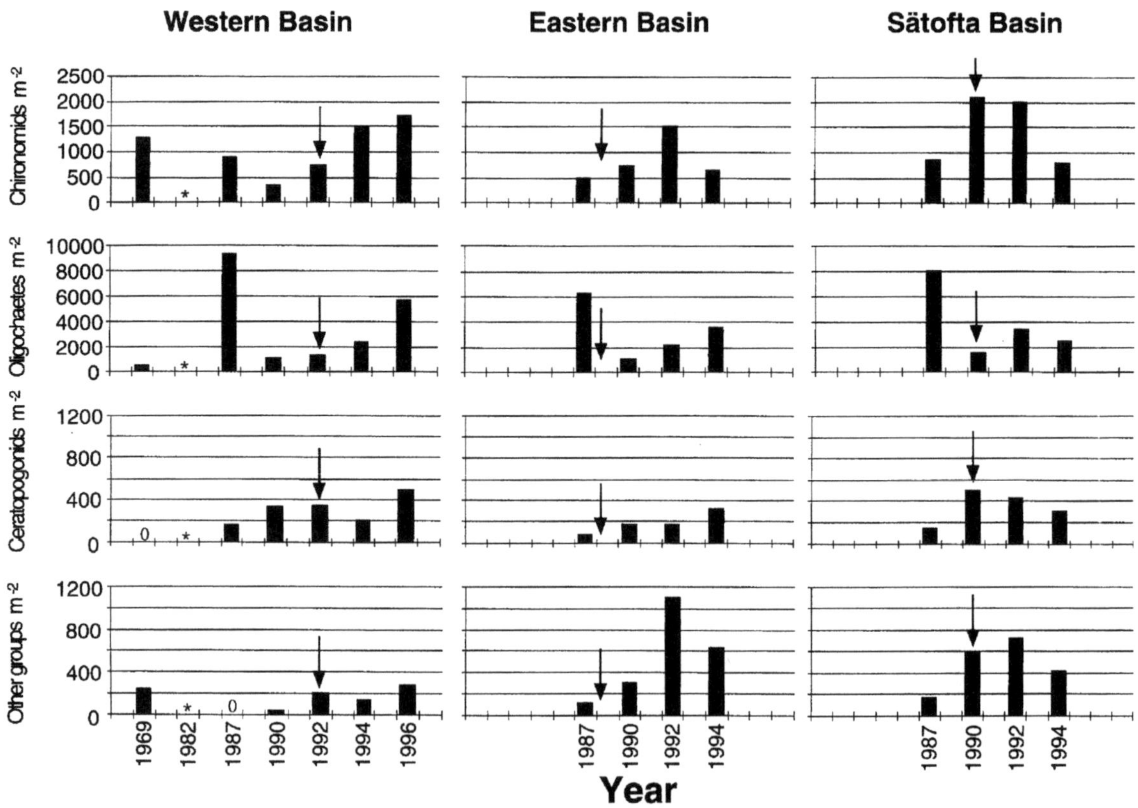

Figure 3. Abundance of chironomids, oligochaetes, ceratopogonids and other groups expressed as depth weighted means over whole lake basin in Western Basin, Eastern Basin and Sätofta Basin. Final period of fish reduction is indicated by arrows. In 1982, fauna was only collected at 1 and 2 meters, which is indicated by '*'.

after fish reduction (Figure 3, Table 2), but did not change in response to fish reduction in Western Basin (Figure 3, Table 2). Based on data from individual depths (Figure 4) oligochaetes decreased at 3, 5 and 7 m in Sätofta Basin (t=4.85 to 5.92, $p \leq$=0.001, df=14), and at 3, 7, 10 and 13 m in Eastern Basin (t=2.85 to 12.4, p <0.001 to $\leq$=0.01, df=19). In Western Basin oligochaete abundance increased at a depth of 1 m (t=2.94, p <0.01, df=17), which was opposite to that found in Sätofta Basin and Eastern Basin.

Ceratopogonids

The first time ceratopogonids were recorded in Western Basin was 1982. They now occur in large numbers in all basins and at almost every depth. Depth-weighted values (ln-transformed, 1987–1996) of abundance show that the ceratopogonids increased in Sätofta Basin and Eastern Basin following the fish reduction, whereas their abundance did not change in Western Basin (Figure 3, Table 2). Analyses for each depth (Figure 4) showed that the ceratopogonid abundance increased at 1, 2 and 13 m (t=2.15 to 2.60, p<0.05, df=14) in Sätofta Basin, at 3 and 5 m (t=2.74 and 3.98, p<0.001 and <0.05, df=19) in Eastern Basin and at 1 and 2 m (t=4.59 and 3.70, p<0.001 and <0.05, df=17) in Western Basin.

Other groups

Nematodes, hirudineans, isopods, amphipods, ephemeropterans, trichopterans, neuropterans, coleopterans, gastropods and lamellibranchs comprised the group, 'other faunal groups'. The abundance of the group 'other' increased after the fish reduction in all basins (Figure 3, Table 2), but only in the shallow areas, i.e.

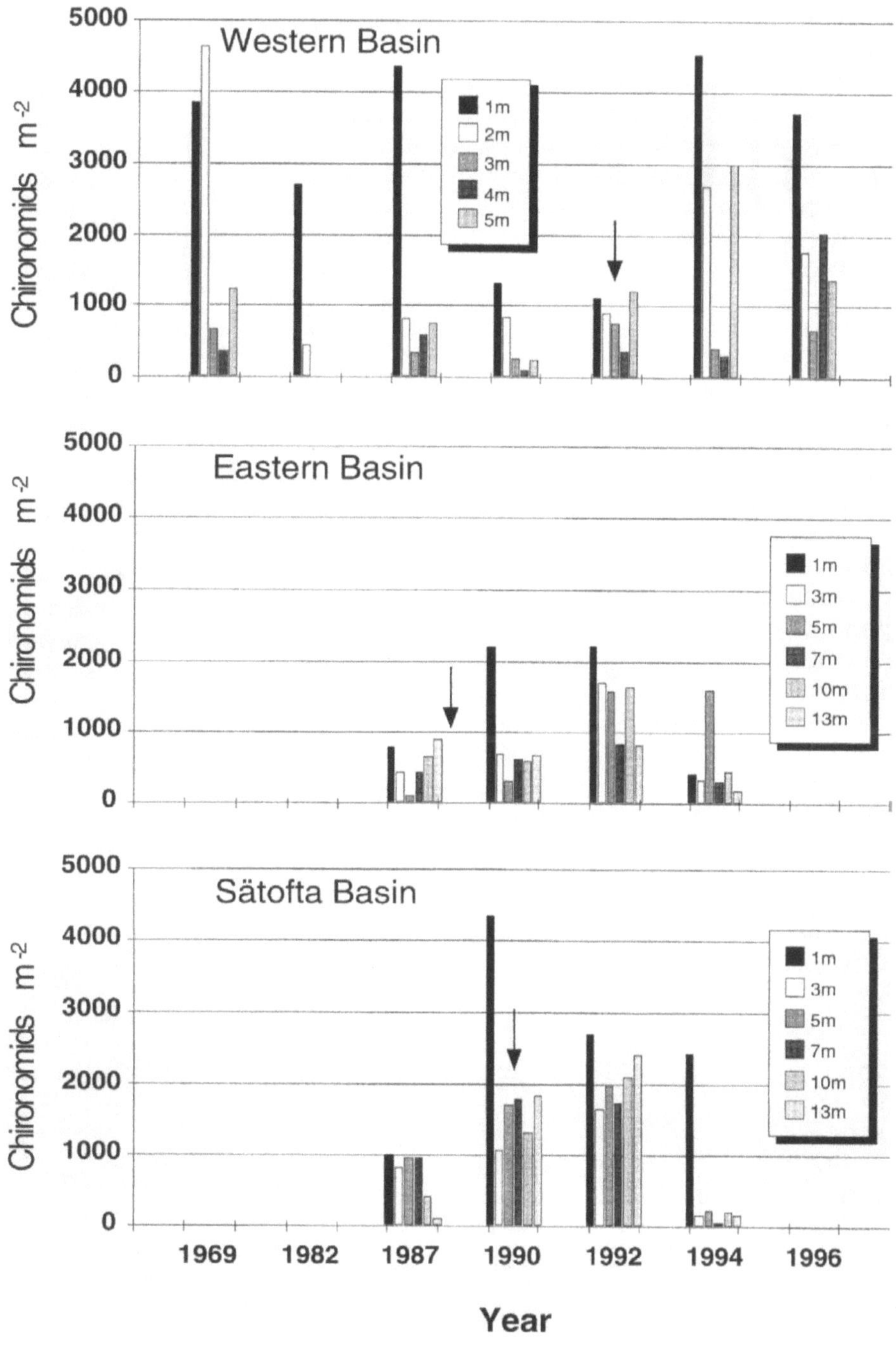

Figure 4a. Abundance (ind. m^{-2}) at each sampled depth of (a) chironomids and (b; next page) other groups in the different basins of Lake Ringsjön. Observe that in Eastern Basin and Sätofta Basin sampling was only performed from 1987 to 1994. Arrows indicate final period of fish reduction.

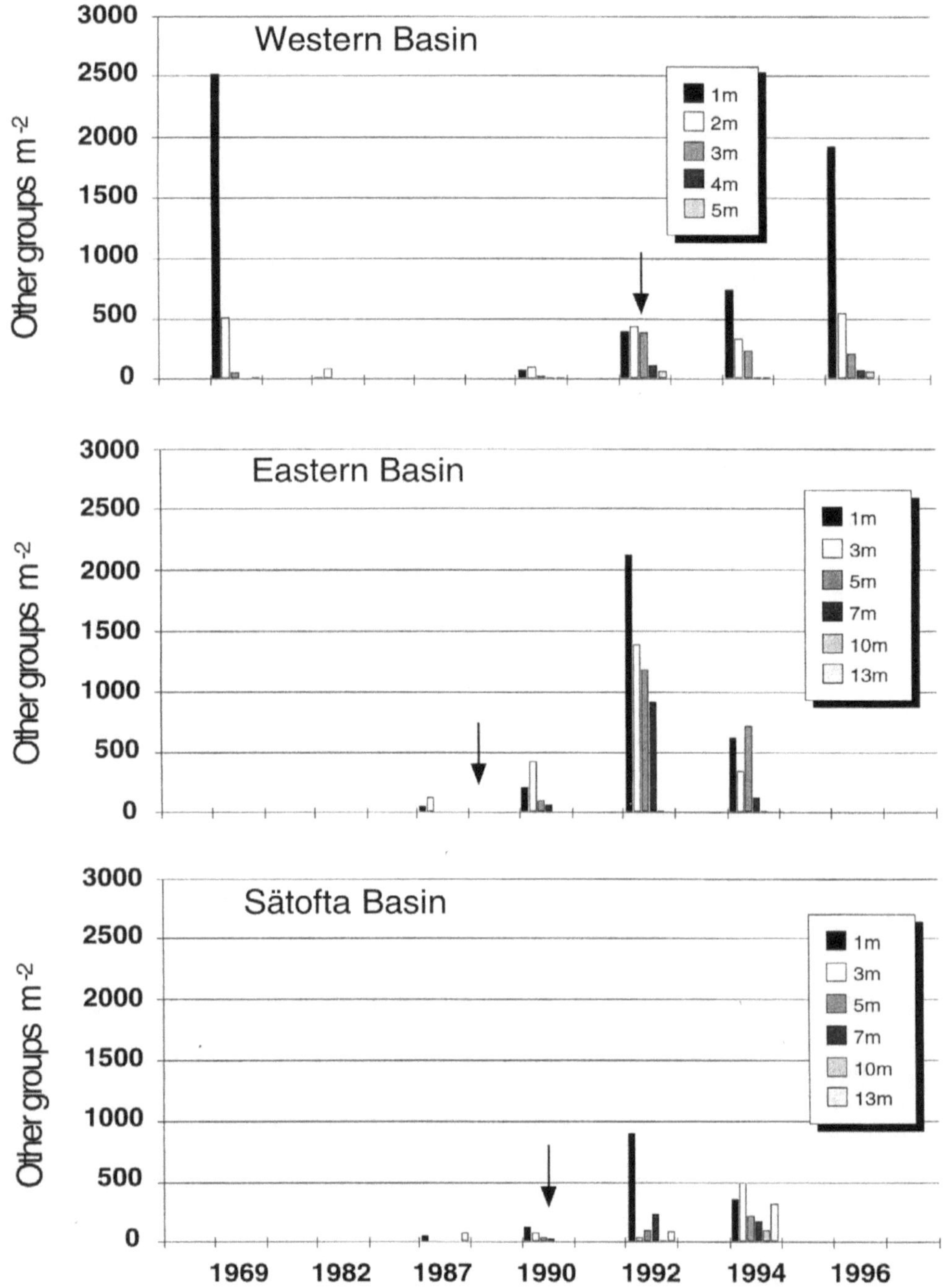
Western Basin
Other groups m -2
3000
2500
2000
1500
1000
500
0
1m
2m
3m
4m
5m
Eastern Basin
Other groups m -2
1m
3m
5m
7m
10m
13m
Sätofta Basin
Other groups m -2
1m
3m
5m
7m
10m
13m
1969
1982
1987
1990
1992
1994
1996

Figure 4b.

at 1 m (t=2.69, p<0.05, df=14) in Sätofta Basin, at 1, 3 and 5 m (t=2.11 to 2.75, p<0.05, df=19) in Eastern Basin and at 1, 2 and 3 m (t=3.20 to 6.34, p<0.01, df=17) in Western Basin (Figure 4).

The Lamellibranchiate, *Pisidium* sp. (Figure 5) had similar abundances in 1969 and 1996, yet was almost eliminated during the years in between. However, it is difficult to draw any firm conclusions here since the sediment collected in 1969 was passed through a 1 mm sieve, probably resulting in an underestimation for 1969 relative to data from 1994 and 1996.

Fish diet

The proportion of chironomids, ceratopogonids and 'others' increased in the diet of bream in 1994 relative to 1987, whereas only the proportion of the group 'other' increased in the diet of roach and perch (Figure 6a). The estimated predation pressure, however, decreased or was the same in 1987 and in 1994 (Figure 6b).

Discussion

Eutrophication effects on the macrofauna community

Generally, a moderate increase in the nutrient supply causes an increase in abundance of macrofaunal groups such as oligochaetes, chironomids, gastropods and sphaerids, and a decrease in ephemeropterans (Jonasson, 1969; Peterka, 1972; Hergerader & Lessig, 1980). If organic enrichment continues and periods of low oxygen, high pH and alkalinity become prolonged, oligochaetes will further increase in abundance. Thus, a macrofauna community dominated by a few species is common in highly eutrophicated waters (Jonasson, 1969). Our results from Western Basin follow this general pattern with a decrease in number of faunal groups during the period 1969–1987 and a predominance of oligochaetes and chironomids in 1982 and 1987. We have no data from 1969 from the two other basins, but the situation was similar in Eastern Basin in 1982 (Cederberg & Runesson, 1982) and in all basins in 1987. Although the external phosphorus load to Lake Ringsjön was much lower in 1984 (c. 10 ton yr^{-1}) compared to the late 1970s (c. 20–30 ton yr^{-1}), macrofaunal community structure in 1987 still reflects a highly eutrophicated situation.

Effects of fish predation on the macrofauna community

Studies on the effects on macrofauna related to fish predation are often contradictory, which probably reflects the many complex interactions occurring between fish predators and different types of macrobenthos. In a whole lake study, Berg et al. (1994) found a decrease in the macrobenthos after introducing whitefish (*Coregonus laveratus* L.), whereas Riemann et al. (1990) found no significant difference in any of the larger taxonomic macrobenthos groups two years after a reduction of roach and bream. In Lake Ringsjön, we found an increase in abundance and biomass of macrobenthos after the fish reduction, which was expected since the removed fish were to a large extent benthivorous bream. Species diversity was also higher in Lake Ringsjön after the fish reduction. In Western Basin, there were almost as many faunal groups in 1996 as in 1943 and 1969. One explanation of the contradictory results in Riemann et al. (1990) and this study could be that initial macrofaunal composition was far more diverse in the Danish lake than in Lake Ringsjön. In Lake Ringsjön the lake bottom was totally dominated by oligochaetes and chironomids, whereas the lake studied by Riemann et al. (1990) also contained species of Mollusca, Crustacea, Odonata and Ephemeroptera. This indicates a low initial predation pressure of fish on benthic macroinvertebrates in Riemanns study, and any removal of fish would not dramatically increase the invertebrate diversity and abundance.

Littoral and profundal effects of the fish reduction

Enclosure studies have shown that roach, bream and perch generally have an effect on macrobenthos; large predacious macrobenthos species decrease in biomass and/or size, whereas for example chironomids might increase in biomass (Andersson et al., 1978; Andersson, 1984; Persson & Greenberg, 1990; Bergman, 1990; Diehl, 1992; Bergman & Greenberg, 1994). The increase of chironomids has been related to fish selecting larger and mobile animals, which include macrobenthic predators. Hence, the reduction of invertebrate predators reduces predation pressure on, for example, chironomids (Kajak, 1988; Diehl, 1992). This is in agreement with our results from Lake Ringsjön where we found an increase in macrofaunal abundance, especially in the group 'other' in the littoral zone, indicating decreased predation from

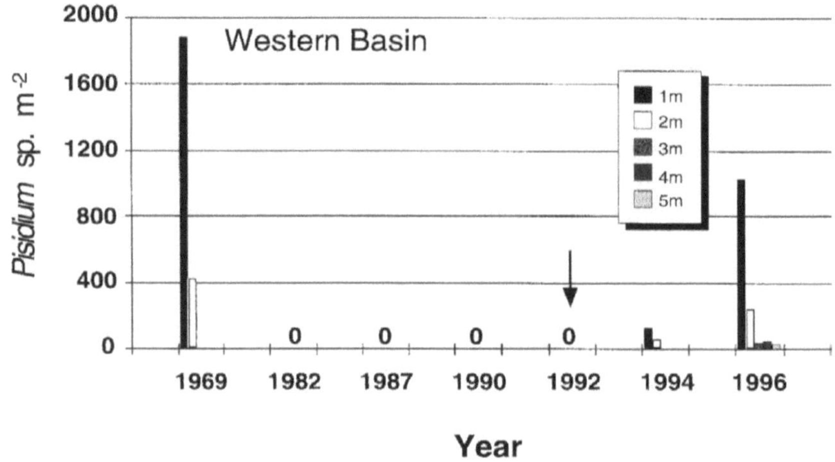

Figure 5. Pisidium abundance (ind. m^{-2}) at each sampled depth in Western Basin. Final period of fish reduction is indicated by arrow.

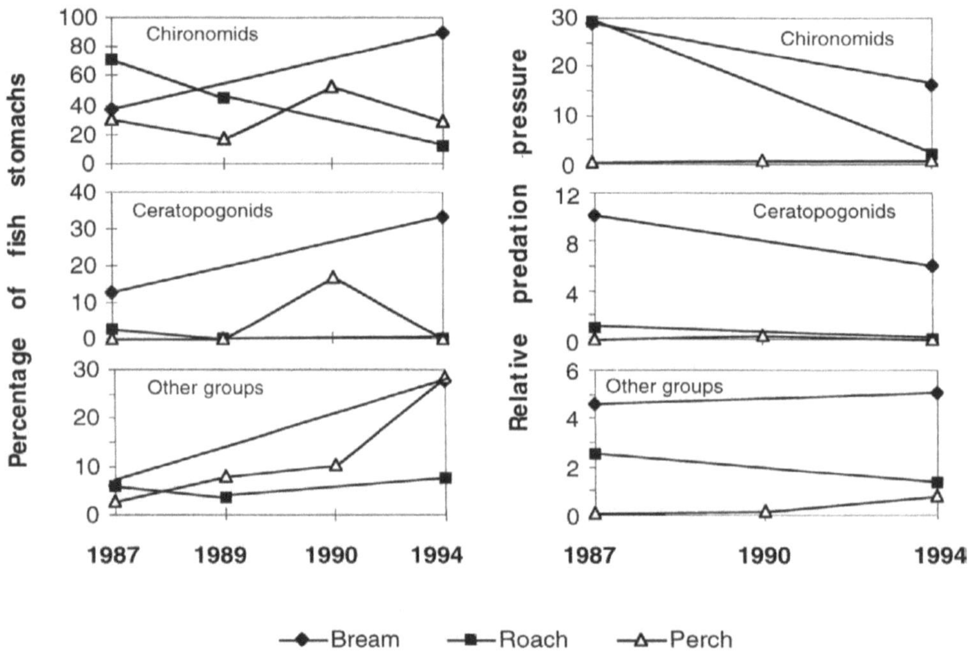

Figure 6. Percentage of fish stomachs with different prey items (a) in 1987, 1989, 1990 and 1994 for perch, roach and bream. The relative predation pressure (b) on different prey items from perch, roach and bream in 1987, 1990 and 1994. See text for further details.

fish. Even the abundance of ceratopogonids has increased most markedly in the littoral zones of the lake basins. Furthermore, the chironomids also increased in abundance in the profundal zones of Western Basin, indicating that predatory macrobenthos might not have strong influence on the benthic communities of larger lakes like Lake Ringsjön.

The sphaeriid clam *Pisidium* sp., made a remarkable recovery in the littoral zone of Western Basin in 1994 and 1996 and had abundances similar to those reported in 1969 (Figure 5). In the eutrophic Danish Lake Ring, Berg et al. (1994) observed that *Pisidium* sp. was the only invertebrate group that increased after introducing whitefish (*C. lavaretus*). This increase in abundance was probably a result of improved food conditions due to reduced competition from other benthic invertebrates. The increase of *Pisidium* sp. in the littoral zone of Lake Ringsjön contradicts the results of Berg et al. (1994), but may relate to the reduction of bream. Bream is a highly effective predator, not only on chironomids but also on *Pisidium* sp. (Muus-Dahlström, 1981). However, in the Lake Ringsjön study we never found *Pisidium* sp. in the stomach content of bream.

Another explanation for the enhanced littoral populations of benthos in Lake Ringsjön, including *Pisidium* sp., could be an overall increase in the abundance of submerged macrophytes, offering shelter from fish predation (Kajak, 1988). However, the few stands of submerged macrophytes in Lake Ringsjön have only marginally increased after the fish reduction (Strand, 1999). The observed increase in the benthic invertebrate abundance in Lake Ringsjön is believed to be caused by the fish reduction. However, it can not be excluded that the simultaneous increase in water transparency of Lake Ringsjön also observed after the fish reduction (Bergman, 1999) may have contributed to this positive development of the benthic fauna. Increased transparency in Lake Ringsjön from about 0.5 m to 1.0 m might not seem dramatic but could presumably provide improved oxygen conditions for the benthos in the littoral zones of the lake due to less sedimentation of organic material.

The role of invertebrates on the recirculation of nutrients from sediment to water

Benthic invertebrates are not only an important food resource for fish and waterfowl, but may also affect nutrient circulation over the sediment–water interface (Petr, 1977; Krantzberg 1985; Andersson et al., 1988). This means that bioturbation, i.e. foraging behavior, construction of burrows and water pumping activity, may contribute to an increased degradation of organic material at the sediment surface. Hence, several studies have shown that animal activity (especially chironomids have been investigated) can stimulate the release of inorganic phosphorus and nitrogen from eutrophic lake sediments to the overlying water (cf. Kitchell et al., 1979; Fukuhara & Sakamoto, 1987, 1988; Andersson et al., 1988). The stimulated release of phosphorus by animal activity must however be considered relative to sediment redox- and pH-conditions that primarily may determine the flux of phosphorus over the sediment–water interface.

Because the fish reduction was associated with an enhanced benthic fauna abundance in Lake Ringsjön, an increase in phosphorus release to the water via increased bioturbation would be expected. Benthic feeding fish such as bream may also recirculate considerable amounts of phosphorus through excretion. Recirculation may also occur as a result of benthic feeding that translocates phosphorus from the sediment to the water and by pelagic feeding recirculating phosphorus within the water column (cf. Andersson et al., 1988; Persson, 1997). Fish-generated phosphorus is probably of greater importance than release caused by invertebrate bioturbation. Phosphorus released from fish in the eutrophic Lake Finjasjön, as calculated by Persson (1997), was of the same order of magnitude as the external load and 42% of the internal load. The fish reduction performed in Lake Ringsjön might presumably increase the mobilisation of phosphorus from the sediment to the water due to increased invertebrate bioturbation after the benthic animal abundance increased. On the other hand, the reduced fish stock should lead to a decreased phosphorus mobilisation by fish excretion and translocation. Hence, the expected net effect would presumably be a lowered phosphorus recruitment from the sediment and phosphorus recirculation within the water column.

Bioturbation activity of tube-dwelling infauna has been shown to increase the rate of bacterial nitrogen transformation processes of nitrification and denitrification in aquatic sediments (Chatarpaul et al., 1979, 1980; Henriksen et al., 1980; Kristensen et al., 1991; Pelegrí et al., 1994; Pelegrí & Blackburn, 1995a,b; Svensson & Leonardson, 1996; Svensson, 1997). In laboratory studies, 2 000 individuals of *Chironomus plumosus* per m^{-2} of sediment surface enhanced denitrification by 2.5–5.6 times compared to sediment without chironomids (Svensson, 1997). These studies were carried out with a eutrophic lake sediment similar to the type found in the profundal zone of Lake Ringsjön, where *C. plumosus* is a common inhabitant. An indication of possibly increased denitrification in Lake Ringsjön is the finding that annual nitrogen retention was significantly greater after the fish reduction. Average nitrogen retention before and after the fish reduction (1984–1987) was 60±6% and 83±14%, respectively. Similar increase in nitrogen retention

Table 3. Annual nitrogen mass balance (tons) for Lake Ringsjön before (1984–1987) and during/after (1988–1996) the fish reduction

Year	Inflow	ΔN within basin	Outflow	Retention	Retention (%)
1984	834	−60	370	528	63
1985	852	23	343	486	57
1986	616	−203	491	328	53
1987	666	−30	259	437	66
1988	897	−210	376	741	83
1989	485	−133	153	465	96
1990	727	−261	169	819	113
1991	800	−30	201	629	79
1992	879	−7	209	677	77
1993	920	−67	283	704	77
1994	985	−7	352	641	65
1995	698	−60	257	501	72
1996	667	−9	87	589	88

capacity was observed by Jeppesen et al. (1998) investigating three eutrophic lakes subjected to major change in fish biomass. The Lake Ringsjön study does however, not include any thorough investigations on phosphorus and nitrogen turnover rates, and direct measurements of denitrification have not been conducted. It is therefore not possible to separate and quantify the effects on nutrient release or retention due to the changes in benthic invertebrate communities in Lake Ringsjön.

In conclusion; the increase in chironomid abundance at the end of the investigation period probably increased food resources for the fish community and may additionally have led to higher nitrification and denitrification rates. Increased denitrification may in turn explain the enhancement in total nitrogen retention in Lake Ringsjön as calculated from mass balance estimates. However, as stated by Jeppesen et al. (1998), there may be several other plausible explanations for increased denitrification beyond enhanced bioturbation. Decreased release of nitrogen due to increased growth of benthic algae and other sediment biota reducing the sediment nitrogen release might, according to Jeppesen et al. (1998), be another explanation. Increased N-retention is considered to be a highly positive effect, leading to decreased transport of nitrogen to the sea.

Acknowledgement

We thank Dr Lars Leonardson for valuable comments on an earlier version of the manuscript, Dr Leif Nilsson and Dr Lars-Anders Hansson for providing data and Dr Larry Greenberg for linguistic corrections. For technical assistance we thank Marie Svensson and Katja Lundin. Financial support was given by the Swedish Fishery Department and the Lake Ringsjön Restoration Committee.

References

Aller, R. C., 1988. Benthic fauna and biogeochemical processes in marine sediments: the role of burrow structures. In T. H. Blackburn & J. Sörensen (eds), Nitrogen Cycling in Coastal Marine Environments. Wiley & Sons, Chichester.

Andersson, G., H. Berggren, G. Cronberg & C. Gelin, 1978. Effects of planktivorous and benthivorous fish on organisms and water chemistry in eutrophic lakes. Hydrobiologia 59: 9–15.

Andersson, G., 1981. Influence of fish on waterfowl and lakes. Anser 20: 21–34 (in Swedish with English summary).

Andersson, G., 1984. The role of fish in lake ecosystems – and in Limnology. In S. Bosheim & M. Nicholls (eds), Interaksjoner Mellom Trofiske Nivåer i Ferskvann. Nordisk limnologsymposium, 3-4 April 1984, Norge. Norsk Limnologforening, Oslo 1984.

Andersson, G., W. Granéli & J. Stenson, 1988. The influence of animals on phosphorus cycling in lake ecosystems. Hydrobiologia 170: 267–284.

Berg, S., E. Jeppesen, M. Søndergaard & E. Mortensen, 1994. Environmental effects of introducing whitefish, *Coregonus laveratus* (L.), in Lake Ring. Hydrobiologia 275/276: 71–79.

Bergman, E., 1990. Effects of roach, *Rutilus rutilus*, on two percids, *Perca fluviatilis* and *Gymnocephalus cernua*: importance of species interaction for diet shifts. Oikos 57: 443–453.

Bergman, E., 1999. Changes in the nutrient load and lake water chemistry in Lake Ringsjön, southern Sweden, from 1966 to 1996. Hydrobiologia 404: 9–18.

Bergman, E. & L. A. Greenberg, 1994. Competition between a planktivore, a bethivore, and a species with ontogenetic diet shifts. Ecology 75: 1233–1245.

Boström, B., M. Jansson & C. Forsberg, 1982. Phosphorus release from lake sediments. Arch. Hydrobiol. Beih. Ergebn. Limnol. 18: 5–59.

Brundin, L., 1949. Chironomiden und andere Bodentieren der Südschwedischen Urbirgsseen. Rept. Inst. Freshw. Res., Drottningholm 30.

Carpenter, S. R. & J. F. Kitchell, 1988. Consumer control of lake productivity. Bioscience 38: 764–769.

Cederberg, E. & M. Runesson, 1982. Bottenfaunan i Östra Ringsjön. (In Swedish). Inst. Report. Dept. Limnology, Lund Univ., 12pp.

Chatarpaul, L., J. B. Robinson & N. K. Kaushik, 1979. Role of tubificid worms on nitrogen transformations in stream sediment. Journal of the Fisheries Research Board of Canada 36: 673–678.

Chatarpaul, L., J. B. Robinson & N. K. Kaushik, 1980. Effects of tubificid worms on denitrification and nitrification in stream sediment. Can. J. Fish. Aquat. Sci. 37: 656–663.

Diehl, S., 1992. Fish predation and benthic community structure: The role of omnivory and habitat complexity. Ecology 73: 1646–1661.

Fukuhara, H. & M. Sakamoto, 1987. Enhancement of inorganic nitrogen and phosphate release from lake sediment by tubificid worms and chironimid larvae. Oikos 48: 312–320.

Fukuhara, H. & M. Sakamoto, 1988. Ecological significance of bioturbation of zoobenthos community in nitrogen release from bottom sediments in a shallow eutrophic lake. Arch. Hydrobiol. 113: 425–445.

Gophen, M., 1990. Biomanipulation: retrospective and future development. Hydrobiologia 200/201: 1–11.

Granéli, W., 1999. Internal phosphorus loading in Lake Ringsjön. Hydrobiologia 404: 19–26.

Hamrin, S. F., 1993. Lake restoration by cyprinid control in Sätofta bay (Lake Ringsjön). Verh. Int. Ver. Limnol. 25: 487–493.

Hamrin, S. F., 1999. Planning and execution of the fish reduction in Lake Ringsjön. Hydrobiologia 404: 59–63.

Hansson, L.-A., M. Enell & E. Bergman, 1999. Lake Ringsjön: its catchment area, its history and its importance. Hydrobiologia 404: 1–7.

Henriksen, K., J. I. Hansen & T. H. Blackburn, 1980. The influence of benthic infauna on exchange rates of inorganic nitrogen between sediment and water. Ophelia Supplementum 1: 249–256.

Hergenrader, G. L. & D. C. Lessig, 1980. Eutrophication of the Salt Valley reservoirs, 1968–73 III. The macroinvertebrate community: Its development, composition, and change in response to eutrophication. Hydrobiologia 75: 7–25.

Hosper, S. H. & E. Jagtman, 1990. Biomanipulation additional to nutrient control for restoration of shallow lakes in the Netherlands. Hydrobiologia 200/201: 523–534.

Jeppesen. E., M. Søndergaard, E. Mortensen, P. Kristensen, B. Riemann, H. J. Jensen, J. P. Müller, O. Sortkjaer, J. P. Jensen, K. Christoffersen, S. Bosselmann & E. Dall, 1990a. Fish manipulation as a lake restoration tool in shallow, eutrophic temperate lakes 1: Cross-analysis of three danish case-studies. Hydrobiologia 200/201: 205–218.

Jeppesen. E., J. P. Jensen, P. Kristensen, M. Søndergaard, E. Mortensen, O. Sortkjaer & K. Olrik, 1990b. Fish manipulation as a lake restoration tool in shallow, eutrophic temperate lakes 2: Threshold levels, long-term stability and conclusions. Hydrobiologia 200/201: 219–227.

Jeppesen. E., J. P. Jensen, M. Søndergaard, T. Lauridsen, P. H. Møller & K. Sandby, 1998. Changes in nitrogen retention in shallow eutrophic lakes following a decline in density of cyprinids. Arch. Hydrobiol. 142: 129–151.

Jonasson, P., 1969. Bottom fauna and eutrophication. In Eutrophication: Causes, Consequences, Correctives. National Acad. Sci., Washington, D.C.: 274–305.

Kajak, Z., 1988. Considerations on benthos abundance in Freshwaters, its factors and mechanisms. Int. Rev. ges. Hydrobiol. 73: 5–19.

Kitchell, J. F., R. V. O'Neill, D. Webb, G. W. Gallepp, S. M. Bartell & J. F. Koonce, 1979. Consumer regulation of nutrient cycling. Bioscience 29: 28–34.

Kranzberg, G., 1985. The influence of bioturbation on physical and biological parameters in aquatic environments; A review. Environ. Pollut. Ser A 39: 99–122.

Kristensen, E., 1988. Benthic fauna and biogeochemical processes in marine sediments: microbial activities and fluxes. In T.H. Blackburn and J. Sörensen (eds), Nitrogen Cycling in Coastal Marine Environments. Wiley & Sons, Chichester.

Kristensen, E., M. H. Jensen & R. C. Aller, 1991. Direct measurement of dissolved inorganic nitrogen exchange and denitrification in individual polychaete (*Nereis virens*) burrows. J. Mar. Res. 49: 355–377.

Lammens, E. H. R. R., H. W. de Nie, J. Vijverberg & W. L. T. van Densen, 1985. Resource partitioning and niche shifts of bream (*Abramis brama*) and eel (*Anguilla anguilla*). Mediated by predation of smelt (*Osmerus eperlanus*) on *Daphnia hyalina*. Can. J. Fish. Aquat. Sci. 42: 1342–1351.

Muus, B. J. & P. Dahlström, 1981. Sötvattenfisk och fiske (In Swedish). P.A. Nordstedt & Söners förlag, Stockholm. ISBN 91-1-814002-5.

Nilsson, L., 1972. Local distribution, food choice and food consumption of diving ducks of a South Swedish lake. Oikos 23: 82–91.

Pelegrí, S. P., L. P Nielsen & T. H. Blackburn, 1994. Denitrification in estuarine sediment stimulated by the irrigation activity of the amphipod *Corophium volutator*. Marine Ecology Progress Series 105: 285–290.

Pelegrí, S. P. & T. H. Blackburn, 1995a. Effect of bioturbation by *Nereis* sp., *Mya arenaria* and *Cerastoderma* sp. on nitrification and denitrification in estuarine sediments. Ophelia 42: 289–299.

Pelegrí, S. P. & T. H. Blackburn, 1995b. Effects of *Tubifex tubifex* (Oligochaeta: Tubificidae) on N-mineralization in freshwater sediments, measured with ^{15}N-isotopes. Aquat. Microbial Ecol. 9: 289–294.

Persson, A., 1997. Fish excretion in relation to internal and external load in an eutrophic lake. Limnol. Oceanogr. 42: 577–583.

Persson, L., 1983a. Food consumption and competition between age classes in a perch (*Perca fluviatilis*) population in a shallow eutrophic lake. Oikos 40: 197–207.

Persson, L., 1983b. Effects of intra- and inter-specific competition on dynamics and size structure of a perch *Perca fluviatilis* and a roach *Rutilus rutilus* population. Oikos 41: 126–132.

Persson, L. & L. Greenberg, 1990. Juvenile competitive bottlenecks: the perch (*Perca fluviatilis*)–roach (*Rutilus rutilus*) interaction. Ecology 71: 44–56.

Persson, L. G., G. Andersson, S. F. Hamrin & L. Johansson, 1988. Predator regulation and primary production along the productivity gradient of temperate lake ecosystems. In S. R. Carpenter (ed.), Complex Interactions in Lake Communities. Springer Verlag, New York.

Peterka, J. J., 1972. Benthic invertebrates in Lake Ashtabula reservoir, North Dakota. Am. Midl. Nat. 88: 408–418.

Petr, T., 1977. Bioturbation and exchange of chemicals in the mud-water interface. In H. L. Golterman (ed.), Interaction Between Sediment and Freshwater. Dr W. Junk, The Hague: 216–226.

Riemann, B., K. Christoffersen, H. J. Jensen, J. P. Müller, C. Lindegaard & S. Bosselmann, 1990. Ecological consequences of a manual reduction of roach and bream in a eutrophic, temperate lake. Hydrobiologia 200/201: 241–250.

Shapiro, J., 1990. Biomanupulation: the next phase – making it stable. Hydrobiologia 200/201: 13–27.

Strand, J., 1999. The development of submerged macrophytes in Lake Ringsjön after biomanipulation. Hydrobiologia 404: 113–121.

Svensson, J. M. & L. Leonardson, 1996. Effects of bioturbation by tube-dwelling chironimid larvae on oxygen uptake and denitrification in eutrophic lake sediments. Freshwat. Biol. 35: 289–300.

Svensson, J. M., 1997. Influence of *Chironomus plumosus* larvae on ammonium flux and denitrification (measured by acetylene blockage- and isotope pairing-technique) in eutrophic lake sediment. Hydrobiologia 346: 157–168.

Svärdson, G., 1976. Interspecific population dominance in fish communities of Scandinavian lakes. Rept. Inst. Freshw. Res., Drottningholm 55: 144–171.

Walls, M., I. Kortelainen & J. Sarvala, 1990. Prey responses to fish predation in freshwater communities. Ann. Zool. Fennici 27: 1–16.

Hydrobiologia **404:** 113–121, 1999.
L.-A. Hansson & E. Bergman (eds), Nutrient Reduction and Biomanipulation as Tools to Improve Water Quality: The Lake Ringsjön Story.

The development of submerged macrophytes in Lake Ringsjön after biomanipulation

John A. Strand
Department of Limnology, Insitute of Ecology, Ecology Building, University of Lund, SE-223 62 Lund, Sweden
Fax: [+46] 462224003; E-mail: john.strand@limnol.lu.se

Key words: distribution, seed bank, water depth, *Potamogeton*, *Myriophyllum*

Abstract

The maximum water depth of submerged vegetation and the number of sites colonized by submerged macrophytes in Lake Ringsjön were studied in 1992, 1993 and 1996, and compared with data from 1947 and 1988, in order to investigate the development of submerged macrophytes after the biomanipulation of the lake (completed 1992). The submerged vegetation has declined considerably since 1947, both in species number and outer water depth. The submerged macrophytes in Lake Ringsjön did not show any clear improvement in outer water depth or number of sites colonized after biomanipulation. The lack of any larger increase in Secchi depth after biomanipulation along with a shortage of suitable habitats (i.e. substrate), waterfowl grazing and species composition of the macrophyte populations are discussed as possible reasons for the poor development of submerged macrophytes in Lake Ringsjön.

Introduction

The importance of submerged vegetation has been increasingly appreciated in biomanipulation projects in accordance with the concept of alternative stable states (van Dijk & van Donk, 1991; Scheffer et al., 1993; Jeppesen et al., 1997). Whether a lake remains in a clear-water state after biomanipulation, or not, seems to a large extent to depend on the development of the submerged vegetation following restoration (Jeppesen et al., 1997). Submerged macrophytes have been claimed to affect the lake ecosystem by a number of mechanisms (Carpenter & Lodge, 1986; Schriver et al., 1995) and are thought to preserve the lake in a clear-water state (Scheffer et al., 1993). Although the mechanisms by which submerged macrophytes affect a lake are not fully understood, various suggestions have been put forward. They include for example, nutrient competition (van Donk et al., 1990), allelopathic substances (Wium-Andersen, 1987) and refugia for herbivorous zooplankton (Timms & Moss, 1984; Beklioglu & Moss, 1996). Furthermore, submerged macrophytes stabilize the sediment and reduce water flow (Carpenter & Lodge, 1986). An increase in submerged macrophytes may also lead to increased periphytic production as the surface available for periphytic algae increases, thus leading to an increased nutrient competition between phytoplankton and periphytic algae.

Important questions to ask before performing biomanipulations are: Can all lakes be restored by biomanipulation? and; can the results be improved by enhancing submerged macrophyte growth? (Scheffer et al., 1993; Jeppesen et al., 1997). An important aspect of these questions is the lake's morphometry. The lake has to have sufficiently large (25–30% of lake area) areas of shallow water (< 2 m) possible for the macrophytes to colonize, once an improvement in water transparency by fish removal or other restoration methods occurs (Hosper & Meijer, 1992). Another factor suggested as important for the re-establishment of submerged macrophytes is existing seed banks (Hosper & Meijer, 1992). Moreover, grazing by birds and fish can delay the development of submerged vegetation after restoration (Lauridsen et al., 1993; Jeppesen et al., 1997). Thus, it is important to investigate the possible conditions for a rapid re-colonization of submerged macrophytes before the restoration project starts, and it is equally important to monitor the development of submerged macrophytes after lake restoration, in order to be able to take appropriate steps for enhancing re-colonization, if necessary.

The aim of this paper is to present data from a monitoring project of submerged macrophyte species after the biomanipulation of Lake Ringsjön. I also present the results of a seed bank investigation of the lake. Furthermore, I will compare the present depth distribution of submerged vegetation in Lake Ringsjön with available historical data.

Study site

Lake Ringsjön is situated in southernmost Sweden (55° 52′ N; 13° 32′ E) and consists of three basins: Sätofta Basin (4.2 km^2, mean depth 3.0 m, maximum depth 17.0 m), Eastern Basin (20.5 km^2, mean depth 6.1 m, maximum depth 16.4 m) and Western Basin (14.8 km^2, mean depth 3.1 m, maximum depth 5.4 m) (Hansson et al., 1999). As many other lakes in urban and agricultural areas, Lake Ringsjön became eutrophicated during the 1960s and 1970s. In the beginning of the 1980s, a *nutrient reduction programme* was initiated, which reduced the external input of nutrients to the lake from phosphorus loadings exceeding 30 tons per year to around 10 tons per year. About ten years later (1989–1990), a *cyprinid fish reduction programme* (biomanipulation) was carried out in Sätofta Basin removing about 100 tons of fish (about 60% of the standing stock). During the winter of 1988, there was a fish-kill in Eastern Basin, eliminating about 500 tons of fish (about 80%) (Hamrin, 1999). The biomanipulation in Western Basin, which was performed in 1992, removed about 120 tons of fish (about 50%).

Methods

Macrophyte and substrate surveys

The macrophyte surveys were performed on 29 September, 1992, 7–8 October, 1993 and 16 September, 1996. The sampling dates in autumn allow for a maximum time of vegetative spread by submerged macrophytes during summer and should thus give the maximum outer depth, whereas maximum biomass and coverage is reached earlier and was not sampled here. Seventy two sites along the shore-line of the lake were qualitatively sampled each year. The distance between each site was 200–500 m. A portable GPS-unit combined with aerial photographs was used in order to revisit the same sites. The sampling was performed by collecting macrophytes with a rake from a boat, beginning just outside the emergent vegetation at each site and moving lakewards. At each site, the inner and outer depth for each submerged species were measured. The lake bottom was examined at each site and noted as to whether it was firm and sandy, soft and muddy, or stony substrate.

To compare the distribution of submerged vegetation after biomanipulation with earlier data, I used data on outer depths from a survey done in 1947 by Almestrand & Lundh (1951) and also unpublished data collected by Irmgard Blindow in 1988. These data are not used in any statistics but are included so as to see general trends in the distribution of vegetation over a longer time period.

Seed bank investigation

The seed banks were sampled in February 1994. At each of the six sites in Western Basin (Figure 2), ten samples were taken with a sediment sampler (ø=7 cm). The top 5 cm of sediment were sampled. The collected sediments were stirred and spread in containers (ø=20 cm, h=15 cm). Each container was filled with aerated tap water, thus covering the sediment with 14 cm of water. The containers were distributed randomly in a greenhouse with an air temperature of 20 °C. Additional light, optimized for plant growth (illumination of 10 000 Lux), was automatically used between 6.00 and 22.00 when external illumination decreased below 17 000 Lux. The sediment samples were checked for germination of seeds every week. Seedlings were removed, counted and determined to species level. After two months the sediment was stirred to bring potentially non-germinated seeds to the sediment surface. The experiment lasted for five months (terminated when no germination had been recorded for four weeks). The sediment was not investigated for non-germinated seeds.

Results

Substrate surveys

The basins in Lake Ringsjön differ substantially in their morphology and substrate at shallow areas (Figure 1). Sätofta Basin has large shallow areas less than 2 m depth (27% of the surface area), which consist largely of sand (only 11% of the shoreline is stony)

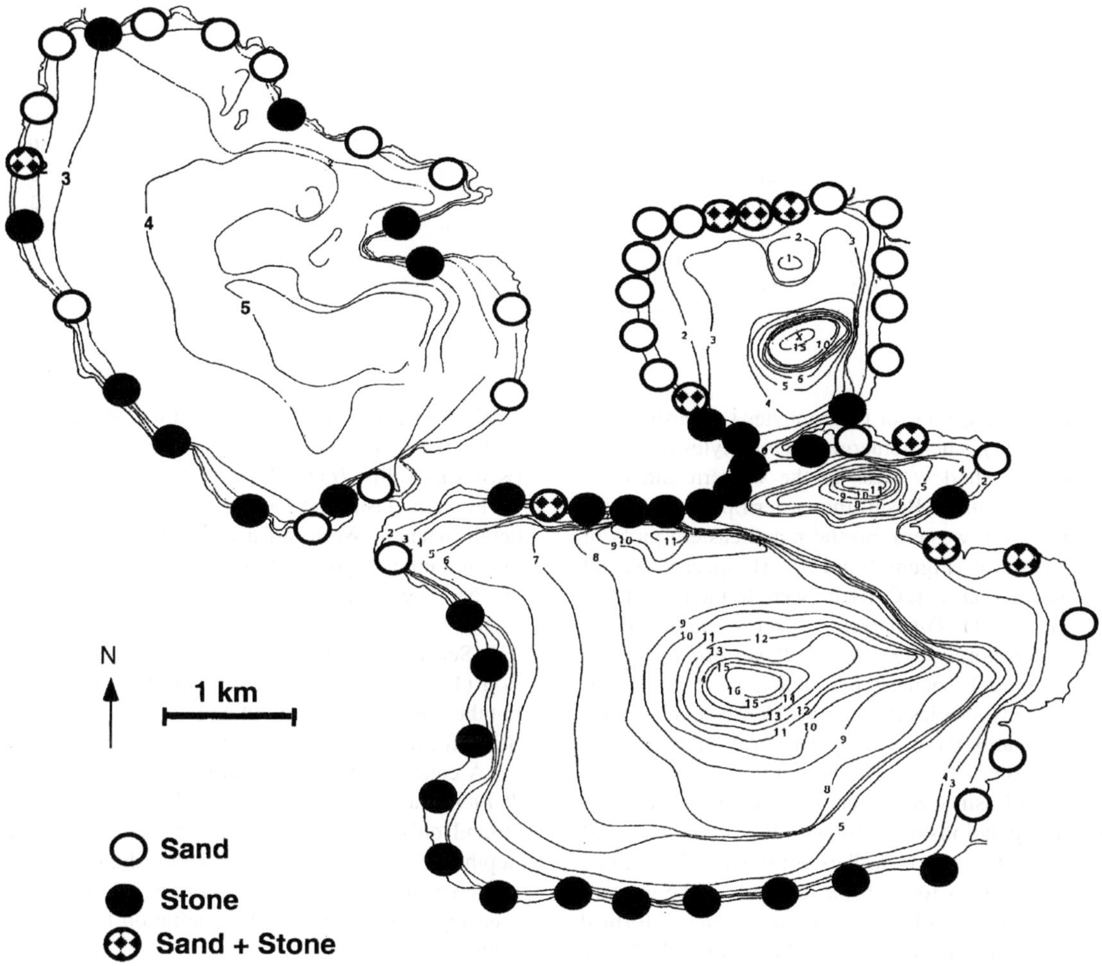

Figure 1. Depth curves and substrate conditions in shallow (<1 m) parts of Lake Ringsjön in the survey of 1992. Stone=more than 50% of lake bottom covered with stone. Sand=more than 50% lake bottom covered with sand. Sand + Stone=approximately 50% of lake bottom was covered with sand and 50% with stone.

(Figure 1). In Western Basin, 16% of the surface area has less than 2 m water depth and a larger part (40%) of the shoreline is stony compared to Sätofta Basin (Figure 1). In Eastern Basin, 11% of the area is less than 2 m deep and 66% of the shoreline is stony. Soft, muddy substrate was not found anywhere at shallow areas, probably due to the large size of the lake which leads to rather high wind exposure.

Macrophyte surveys

A total of 6 submerged angiosperm species were found during the surveys, and no Charophytes (Table 1 and 2). In the 1947 survey, 30 species were found in-

Table 1. Percent of the 72 investigated sites in Lake Ringsjön with submerged macrophytes

Species	Year		
	1992	1993	1996
P. pectinatus	12.5	11.1	12.5
P. crispus	19.4	13.8	23.6
P. perfoliatus	18.1	9.7	15.3
P. lucens	4.2	2.7	6.9
M. spicatum	11.1	4.2	8.3
E. canadensis	2.7	0	2.7
Submerged vegetation	43.1	30.1	43.1

Table 2. Percent of the investigated sites in Sätofta Basin (n=18), Eastern Basin (n=32) and Western Basin (n=22) with submerged macrophytes

	Sätofta			Eastern			Western		
Species:Year	1992	1993	1996	1992	1993	1996	1992	1993	1996
P. pectinatus	11	6	11	19	19	22	9	0	0
P. crispus	56	33	61	6	3	3	14	18	50
P. perfoliatus	28	22	33	19	9	13	5	0	0
P. lucens	0	0	0	0	0	0	14	9	23
M. spicatum	6	0	0	3	0	0	36	9	32
E. canadensis	6	0	6	0	0	0	5	0	5
Submerged vegetation	61	44	56	28	22	22	55	36	64

cluding eight species of Charophytes (Almestrand & Lundh, 1951). The submerged macrophytes are today confined to shallow areas in the western and eastern part of Sätofta Basin, the eastern part of Eastern Basin and the western, northern and eastern parts of Western Basin (Figure 2 and 3). All species, except *Elodea canadensis* L.C. Rich., were found in all three years (Table 1). *Potamogeton* species dominated the submerged flora, *P. crispus* L., *P. perfoliatus* L. and *P. pectinatus* L. being most abundant. *P. pectinatus* was the only species that formed dense stands covering reasonably large areas, particularly in the eastern part of Eastern Basin. The western and eastern parts of Sätofta Basin were vegetated with sparse stands of mixed populations of *P. crispus*, *P. pectinatus* and *P. perfoliatus* (Figure 3). In Western Basin, *M. spicatum* L. and *P. crispus* formed sparse stands at the western and eastern parts of the basin, and *P. lucens* formed dense but small stands at the northern part of the basin.

The water level fluctuations differed substantially between years (Figure 4). The water level at sampling occasions (late summer/autumn) was lowest in 1992 and highest in 1993 (Figure 4). The water levels of 1988, 1992 and 1993 had a maximum in February–March and decreased during spring. Furthermore, the water level of 1993 had a lower amplitude compared to 1988, 1992 and 1996 (Figure 4). In the spring of 1996, the water level was very low with a maximum in May–June. In order to neutralize the effect of different water levels at sampling occasions, all data of outer depth for the vegetation were re-calculated in relation to the water level at the sampling of 1993.

The outer depth (corrected for water level difference on sampling dates) at sites that had submerged vegetation all three years (see Figure 2) did not show any inter-annual variations except for Sätofta Basin where the outer depth was significantly lower in 1993 (0.86 ± 0.33, n=7) compared to 1992 and 1996 (1.32 ± 0.29, n=7 and 1.2 ± 0.28, n=7, respectively) (Kruskal-Wallis, H=6.19, p=0.045).

Eastern Basin had the lowest average outer water depth, whereas Western Basin had the lowest average Secchi depth and highest outer vegetation depth (Figure 5). The higher outer depth in 1947 corresponds to the higher Secchi depth at this time (Figure 5). The Secchi depth was higher after the biomanipulation (1992, 1993, 1996) compared to before (1988), but there was no corresponding increase in outer water depth for the vegetation (Figure 5).

A clear difference existed in outer depth between *P. lucens* and the other species (Figure 6). *P. lucens* (found only in Western Basin) grew at the greatest depths, followed by *P. crispus* (both showing a slight tendency to decrease). *P. pectinatus* showed an increasing tendency in depth penetration from 1992 to 1996 (Figure 6).

Seed bank investigation

No submersed plants germinated from seeds. Four plants originating from vegetative parts (turions) of *P. crispus* were the only submersed macrophytes to 'germinate'.

Discussion

Based on my samples from September to October 1992, 1993 and 1996, and a comparison with outer depth data from 1988, the submerged vegetation in Lake Ringsjön did not respond to the biomanipulation in any obvious way. There were no differences in outer water depth between years, except a tendency towards a decrease in 1993 both as to outer depth,

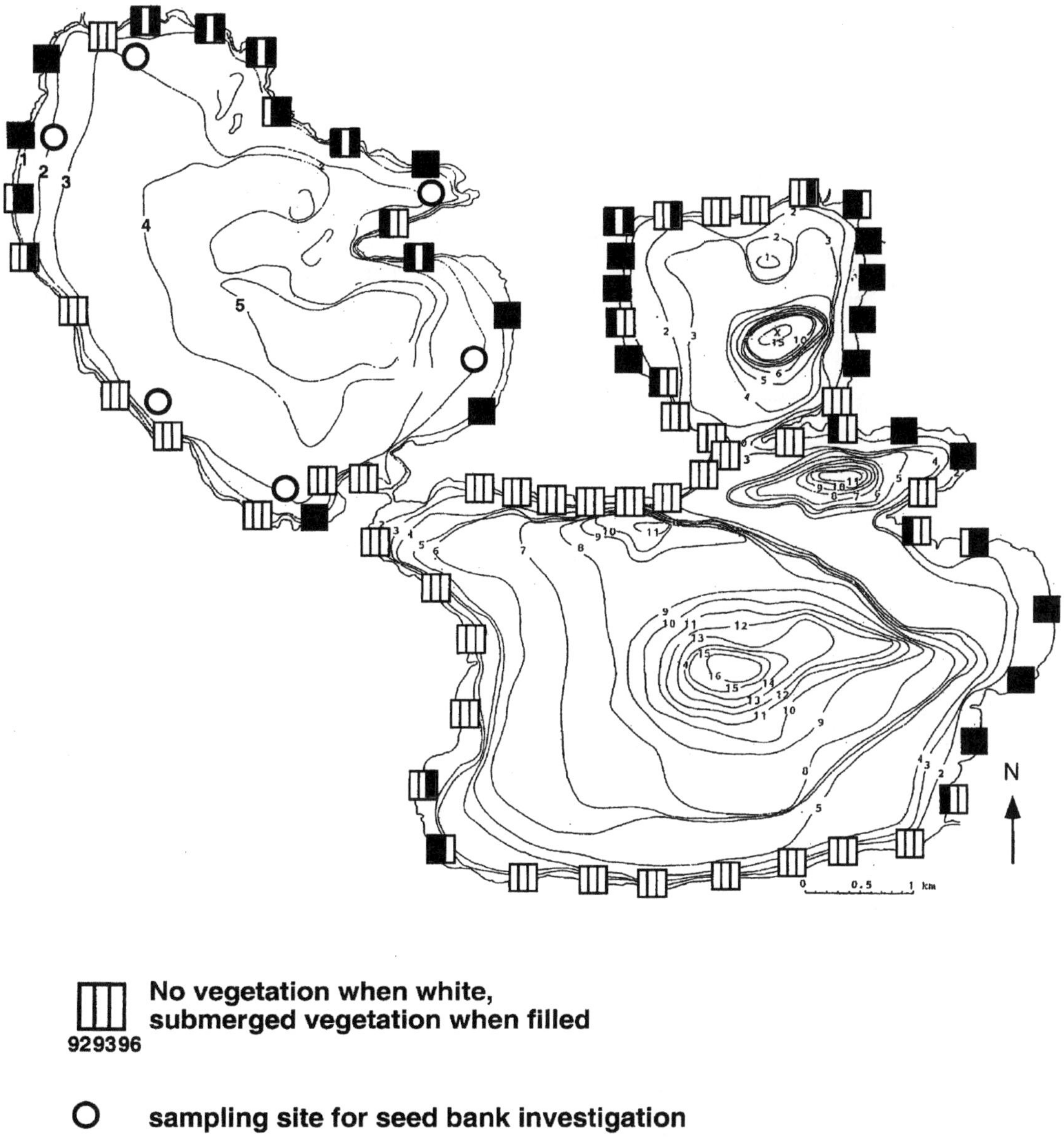

Figure 2. Presence and absence of submersed vegetation at the 72 sites in 1992 (sampled 29 September), 1993 (sampled 7–8 October), 1996 (sampled 16 September).

and number of sites colonized. Although the effect of different water levels at sampling occasions was removed, the effect of different water levels during the vegetation periods may affect the depth distribution at the end of the growing season. The persistent high water level during summer in 1993 may be a reason for the reduced growth at larger depths that year. The historical data available on submerged vegetation indicate a marked decrease in diversity and outer depth over the last 50 years. Almestrand & Lundh (1951) reported additional 24 submerged species, including *Isoetes lacustris* L., *Littorella uniflora* L. and *Myriophyllum alterniflorum* L., which are typical for oligotrophic lakes. Furthermore, they reported dense mats

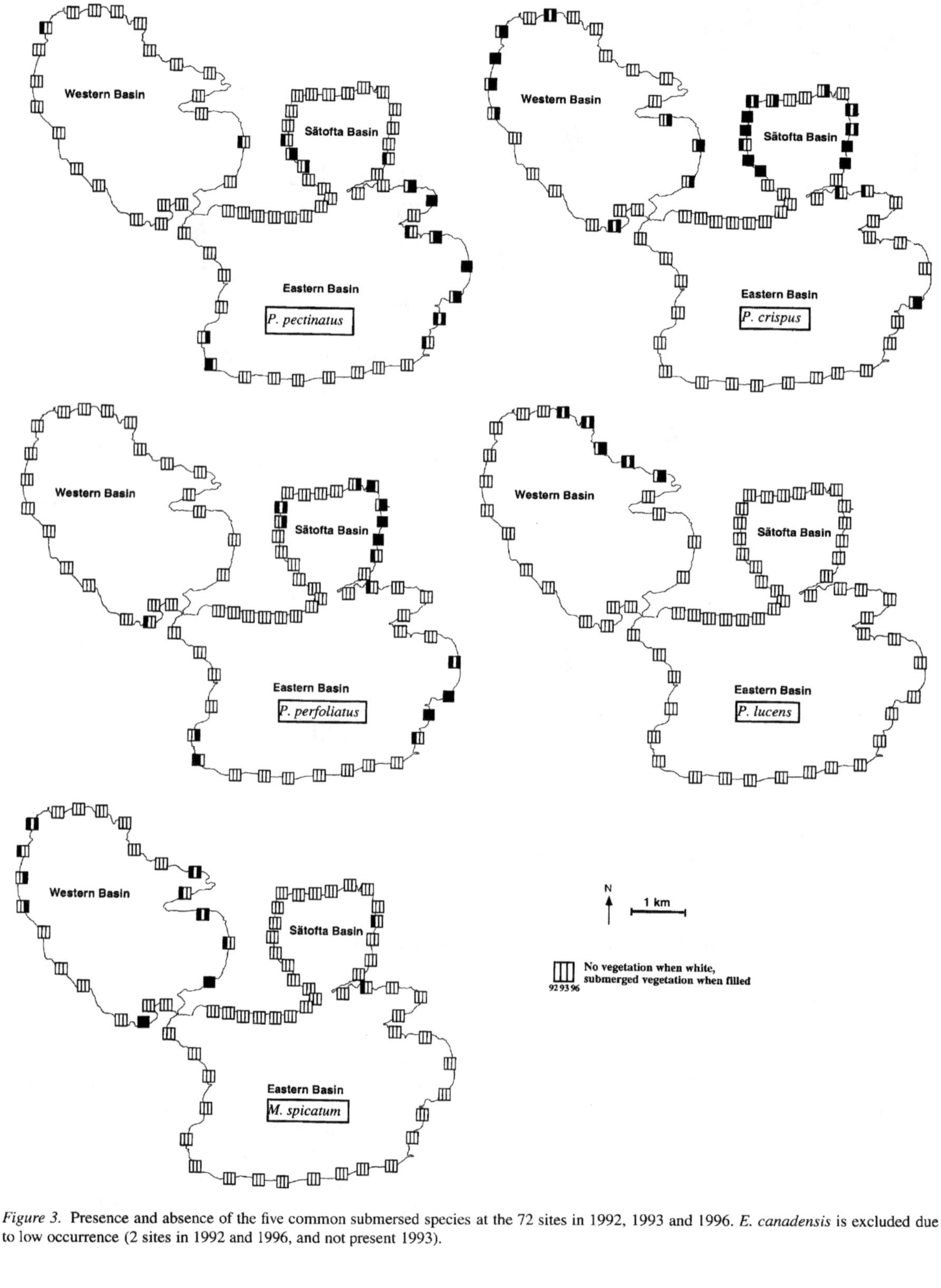

Figure 3. Presence and absence of the five common submersed species at the 72 sites in 1992, 1993 and 1996. *E. canadensis* is excluded due to low occurrence (2 sites in 1992 and 1996, and not present 1993).

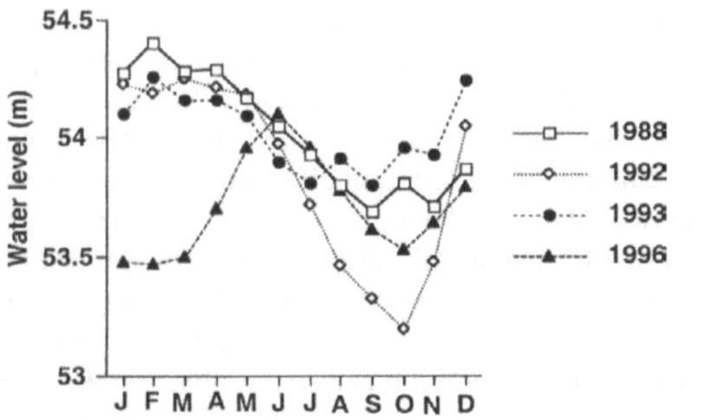

Figure 4. Water level fluctuations during 1988, 1992, 1993 and 1996. The water level is measured at the outflow of Western Basin.

of clearwater-demanding rosette species and Charophytes (8 species) at considerable water depths (2 m). The disappearance of these species indicates a deteriorating light climate leading to the impoverished submerged macrophyte populations found today. In 1996, solitary *Chara globularis* were observed in shallow water in the eastern part of Eastern Basin (H. Sandsten, pers. com.). This is a good sign, but the overall impression following the 3-year investigation (conducted over 4 years) is that submerged vegetation is still very sparse and shows low species diversity.

The poor development of submerged macrophytes after the restoration is in accordance with many other biological and chemical parameters in the pelagial of the lake (Bergman et al., 1999; Bergman & Bergstrand, 1999). An interesting exception to the overall poor response of Lake Ringsjön to biomanipulation, is the development of herbivorous waterfowl populations which increased after the fish removal (see Andersson & Nilsson, 1999). The increase in waterfowl density can be an explanation of the poor development in submerged vegetation, since it has been shown that grazing birds can delay establishment of submerged macrophytes after restoration projects (Lauridsen et al., 1993).

Even though Secchi depth was highest in Eastern Basin, the outer water depth was lowest here. The low outer depth for submerged vegetation in this basin could be explained by a lack of suitable habitats. A large part of the littoral zone is very stony without any substrate in which plants may grow. Sandy bottoms do not appear until a water depth of about 2 m. This was particularly true for the northern and southern shores of Eastern Basin. The high transparency was probably not yet high enough for submerged macrophytes to establish at these sites. Historical data support the no-

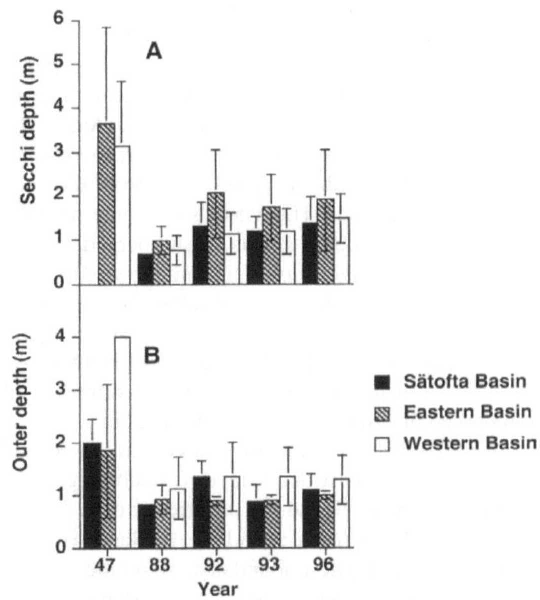

Figure 5. (A) Mean Secchi depth (April–October) for the different basins in 1947 (n=34), 1988 (n=7), 1992 (n=13), 1993 (n=14) and 1996 (n=7). Thin vertical bars represent standard deviation except for 1947 where they represent the maximum value for the period. No value exists for Sätofta Basin in 1947. (B) Mean maximum depth distribution of submerged vegetation in the three basins in Lake Ringsjön in 1947 (n=1–5), 1988 (n=1–3), 1992 (n=9-12), 1993 (n=7–8) and 1996 (n=7–14). Thin vertical bars represent standard deviation.

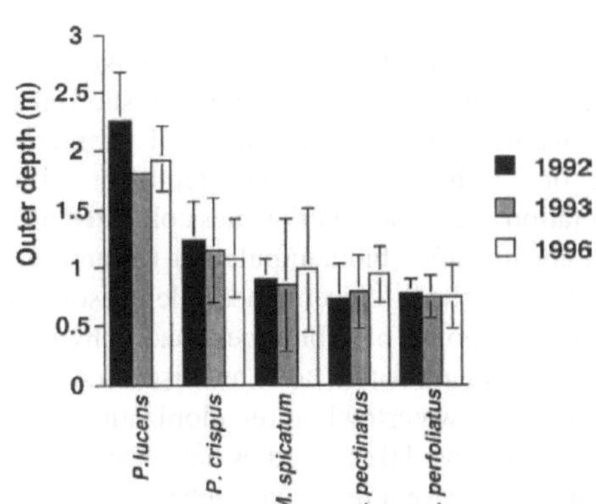

Figure 6. Mean outer depth for the submerged species in 1992, 1993 and 1996 in all basins. Thin vertical bars represent standard deviation.

tion that Eastern Basin has always been the basin with lowest abundance of submerged vegetation (Trybom, 1893; Almestrand & Lundh, 1951). When comparing substrate (Figure 1) with occurrence of vegetation at the sampling sites (Figure 2), it is clear that vegetation only occurs on sandy substrate. Along the eastern shore of Eastern Basin, however, there were suitable habitats with sandy substrate from the reed stands and lakewards. This was also the area with the largest and most dense stands of submerged macrophytes (mainly *P. pectinatus* and some *P. perfoliatus*). The reason for the relatively low values of outer depth at this area was probably a very steep slope beginning at approximately 1.1–1.2 m depth in the two bays of the eastern part of the basin. The distribution of submerged macrophytes stops abruptly here.

Western Basin, on the other hand, had low Secchi depth but the highest outer depth for submerged vegetation, probably due to the occurrence of *P. lucens*. Historically, Western Basin and Sätofta Basin were the basins with the highest abundance of submerged vegetation, probably due to large areas of shallow water and suitable habitats (16 and 27% respectively, of the surface area < 2 m water depth and consists mostly of sandy substrate). In Eastern Basin, only 11% of the surface area had a water depth of 2 m or less, and were furthermore mostly stony, unsuitable habitats. The number of sites with submerged vegetation in the different basins also showed that Sätofta Basin (55% of the sites were vegetated in 1996) and Western Basin (64% of the sites were vegetated in 1996) had better conditions for submerged vegetation compared to Eastern Basin (22% of the sites vegetated), despite higher transparency in Eastern Basin.

Compared to Lake Ringsjön, Lake Finjasjön, which has also been biomanipulated, showed a clear improvement in the development of submerged vegetation. Both the outer water depth for submerged vegetation and the number of sites colonized increased dramatically after biomanipulation (Strand & Weisner, unpublished). Several possible reasons for the different response of submerged macrophytes to biomanipulation in the two lakes are discussed below.

Effects of waterfowl on re-colonization may differ between the two lakes. Studies have shown that bird-grazing can prevent or at least delay re-establishment of submerged macrophytes after biomanipulation projects (Lauridsen et al., 1993). Grazing experiments conducted in Lake Finjasjön in 1992 and 1996 revealed no effect of bird-grazing (and/or fish) on submerged macrophytes (Strand & Weisner, unpublished data), whereas grazing experiments in Eastern Basin showed a negative effect of bird-grazing on *P. pectinatus* both on above-ground and below-ground biomass (H. Sandsten, unpublished data). The simple explanation of the different effects of grazing between the two lakes may be higher densities of grazing birds in Lake Ringsjön, this suggestion is strengthened by the high numbers of swans and coots grazing on submerged macrophytes in Lake Ringsjön, (Andersson & Nilsson, this volume). It is also possible that the palatability of the dominant species differed between lakes. *M. spicatum*, which has been shown to not be sensitive to grazing, dominated in Lake Finjasjön (Weisner et al., 1997), whereas *P. pectinatus*, which is strongly negatively affected by bird-grazing, is common in Lake Ringsjön (Weisner et al., 1997).

The absence of a viable seed bank is not detrimental to successful re-colonization of submerged macrophytes. Seed bank studies in Lake Finjasjön (same approach as the seed bank studies in Western Basin) showed the same absence of seeds as in Western Basin. It is therefore almost certain that the rapid spread of submerged macrophytes in Lake Finjasjön is not caused by germinating seeds. It could be argued that the right conditions for germination had not been obtained and I did not check the sediment for viable, not germinated, seeds. This seed bank investigation was however, part of a larger investigation involving other lakes, where germination of both Charophytes and angiosperms were abundant.

Another explanation for different responses in Lake Finjasjön and Lake Ringsjön may be the species composition of the remaining populations of submerged macrophytes, which are characteristic for eutrophicated lakes. In Lake Ringsjön, the submerged vegetation was dominated by *Potamogeton* spp., whereas in Lake Finjasjön, other elodeid species dominated (*M. spicatum* and *E. canadensis*). *M. spicatum* and *E. canadensis* are more rapidly expanding species compared to *Potamogeton* species. One reason for this is probably the mechanism of fragmentation that *M. spicatum* and *E. canadensis* possess. A small fragment of the stem of these two species may, when dislodged, float in the water for months. The fragments may then establish in shallow water in other parts of the lake (Kimbel, 1982). *M. spicatum* even develops roots from the stem to prepare the fragments and improve the possibilities of successful establishment (Kimbel, 1982). This mechanism allows these species to spread to new areas of a lake during one growing season (Kimbel, 1982).

Species that are dependent on seed dispersal have lower dispersal rates since inherent dormancy of the seeds requires winter cooling for germination to occur (Haag, 1983). Dormancy in fragments of *M. spicatum* and *E. canadensis* however, seems not to exist or to be weak (Haag, 1979; Kimbel, 1982).

It has been suggested that at least 25–30% of the lake area should be re-colonized with submerged vegetation if restoration is to be successful (Hosper & Meijer, 1992). To reach that level in Lake Ringsjön, dense stands of submerged vegetation must expand to 2 m depth in Sätofta Basin, 4 m in Eastern Basin and 3 m in Western Basin. It is obvious that this has not been the case. However, in view of the successful restoration of Lake Finjasjön, at least Western Basin and Sätofta Basin have the potential to be colonized by submerged macrophytes if the fish removal is done more efficiently.

Acknowledgement

Irmgard Blindow kindly let me use her data from 1988. She also did the species determination on the preserved Charophytes from the survey of 1947. Bálint Wagner and Catrine Johansson helped with the field work. Lars-Anders Hansson, Eva Bergman and two anonymous reviewers gave valuable comments on an earlier version of the manuscript.

References

Almestrand, A. & A. Lundh, 1951. Studies on the vegetation and hydrochemistry of scanian lakes 1–2. Botaniska notiser, suppl. 2:3. C. W. K. Gleerups förlag, Lund, Sweden.

Andersson, G. & L. Nilsson, 1999. Autumn waterfowl abundance in Lake Ringsjön. Hydrobiologia 404: 41–51.

Beklioglu, M. & B. Moss, 1996. Existence of a macrophyte-dominated clear water state over a very wide range of nutrient concentrations in a small shallow lake. Hydrobiologia 337: 93–106.

Bergman, E. & E. Bergstrand, 1999. Lack of a top-down effect on the zooplankton community after a cyprinid fish reduction. Hydrobiologia 404: 77-87.

Bergman, E., 1999. Changes in the nutrient load and lake water chemistry in Lake Ringsjön, southern Sweden, from 1966 to 1996. Hydrobiologia 404: 9–18.

Carpenter, S. R. & D. M. Lodge, 1986. Effects of submersed macrophytes on ecosystem processes. Aquat. Bot. 26: 341–370.

Haag, R. W., 1979. The ecological significance of dormancy in some rooted aquatic plants. J. Ecol. 67: 727–738.

Haag, R. W., 1983. Emergence of seedlings of aquatic macrophytes from lake sediments. Can. J. Bot. 61: 148–156.

Hamrin, S. F., 1993. Lake restoration by cyprinid control in Sätofta Bay (Lake Ringsjön). Verh. int. Ver. Limnol. 25: 487–493.

Hamrin, S. F., 1999. Planning and execution of the fish reduction in Lake Ringsjön. Hydrobiologia 404: 59–63.

Hansson, L.-A., M. Enell & E. Bergman, 1999. Lake Ringsjön: its catchment area, its history and its importance. Hydrobiologia 404: 1–7.

Hosper, H. & M.-L. Meijer, 1992. Biomanipulation, will it work for your lake? Simple test for the assessment of chances for clear water following drastic fish-stock reduction in shallow, eutrophic lakes. INTECOL's 4th International Wetlands Conference, Columbus, Ohio, U.S.A.

Jeppesen, E., J. P. Jensen, M. Søndergaard, T. Lauridsen, L. J. Pedersen & L. Jensen, 1997. Top-down control in freshwater lakes: the role of fish, submerged macrophytes and water depth. Hydrobiologia 342/343: 151-164.

Kimbel, J.C., 1982. Factors influencing potential intralake colonization by *Myriophyllum spicatum* L. Aquat. Bot. 14: 295–307.

Lauridsen, T. L., E. Jeppesen & F. Østergaard Andersen, 1993. Colonization of submerged macrophytes in shallow fish manipulated Lake Væng: impact of sediment composition and waterfowl grazing. Aquat. Bot. 46: 1–15.

Scheffer, M., S. H. Hosper, M.-L. Meijer, B. Moss & E. Jeppesen, 1993. Alternative Equilibria in shallow lakes. TREE 8: 275–279.

Schriver, P., J. Bøgestrand, E. Jeppesen & M. Søndergaard, 1995. Impact of submerged macrophytes on fish-zooplankton-phytoplankton interactions: large-scale enclosure experiments in a shallow eutrophic lake. Freshwat. Biol. 33: 255–270.

Timms, D. & B. Moss, 1984. Prevention of growth of potentially dense phytoplankton populations by zooplankton grazing, in the presence of zooplanktivorous fish, in a shallow eutrophic lake. Limnol. Oceanogr. 29: 472–486

Trybom, F., 1893. Ringsjön i Malmöhus län; dess naturförhållanden och fiske. Medd. Kongl. Lantbruksstyrelsen. 4: 3–36 (In Swedish).

van Dijk, G. M. & E. van Donk, 1991. Perspectives for submersed macrophytes in shallow lake restoration projects in the Netherlands. Hydrobiol. Bull. 24: 125–131.

van Donk, E., M. P. Grimm, M. P. Gulati & J. P. G. Klein Breteler, 1990. Whole-lake food web manipulation as a means to study community interactions in a small ecosystem. Hydrobiologia 200/201: 275–289.

Weisner, S. E. B., J. A. Strand & H. Sandsten, 1997. Mechanisms regulating abundance of submerged vegetation in shallow eutrophic lakes. Oecologia, 109: 592–59.

Wium-Andersen, S., 1987. Allelopathy among aquatic plants. Arch. Hydrobiol. Bull. 18: 157–158.

Hydrobiologia **404:** 123–129, 1999.
L.-A. Hansson & E. Bergman (eds), Nutrient Reduction and Biomanipulation as Tools to Improve Water Quality: The Lake Ringsjön Story.

The occurrence of toxic blue-green algae in Lake Ringsjön, southern Sweden, despite nutrient reduction and fish biomanipulation

Gertrud Cronberg[1], Heléne Annadotter[1] & Linda A. Lawton[2]
[1]*Department of Limnology, Insitute of Ecology, Ecology Building, University of Lund, SE-223 62 Lund, Sweden*
[2]*School of Applied Sciences, Robert Gordon University, Aberdeen, Scotland, U.K.*

Key words: Anabaena, Aphanizomenon, Microcystis, Woronichinia naegeliana, biomanipulation, blue-green algae, cyanobacteria, cyanoprokaryotes, hepatotoxin, microcystin

Abstract

Algal nuisance blooms have occurred during summers for more than 40 years in Lake Ringsjön, southern Sweden. Cattle deaths were recorded in connection with a high biomass of blue-green algae as early as 1968. A mouse bioassay in the 1980s confirmed the occurrence of hepatotoxic algae. In 1995, phytoplankton biomass and microcystin concentrations were measured. A mouse bioassay was carried out simultaneously showing that from the end of June, when the blue-green algae appeared, and until their disappearance at the end of October, the blooms were hepatotoxic. The blue-green alga *Microcystis wesenbergii* dominated over most of this period, although, *Microcystis botrys, M. flos-aquae, M. viridis, Aphanizomenon klebahnii* and *Woronichinia naegeliana* also occurred frequently. However, the algal toxicity was not related with any specific alga or with any of the measured parameters. The highest toxin production was measured in July, although the blue-green algal biomass was low at that time.

Introduction

Blue-green algae (cyanobacteria, cyanoprokaryotes) are photosynthetic prokaryotes of which some genera produce potent toxins. In 1878, the first scientific study on a toxin-producing species, *Nodularia spumigena*, MERTENS, poisoning animals was published (Francis, 1878). A large number of later publications report blue-green algal poisoning of cattle (e.g. Dillenberg & Dehnel, 1960), horses (e.g. Schwimmer & Schwimmer, 1964), sheep (e.g. Main et al., 1977) and dogs (e.g. Lundberg et al., 1983). Algal poisoning of wild animals has been reported on birds (e.g. Rose, 1953), fish (e.g. Skulberg, 1988), bats (Pybus & Hobson, 1986), muskrats (Rose, 1953) and mink (Rose, 1953). Human reactions associated with blue-green algae include symptoms such as skin-rash (Cohen & Reif, 1953), fever (Dillenberg & Dehnel, 1960), hay-fever-like symptoms (Heise, 1949, 1951), gastroenteritis (e.g. Dillenberg & Dehnel, 1960) and toxic hepatitis (Falconer et al., 1983; Hawkins et al., 1985). Between February and April 1996, the worst incident of blue-green algal poisoning in humans ever recorded occurred in Brazil (Pouria et al., 1998). Sixty dialysis patients died of acute liver damage caused by a contaminated dialysis fluid. The dialysis fluid was prepared from drinking water contaminated with the blue-green algal toxin Microcystin-LR, a toxin produced by the blue-green alga *Microcystis aeruginosa*. Hepatotoxins, neurotoxins and lipopolysaccharide endotoxins are the main groups of blue-green algal toxins. The toxins are secondary metabolites and are produced intracellularly (Carmichael, 1992). The ecological significance of blue-green algal toxin-production is, however, not yet clarified.

At the end of the 1940s, the water of lake Ringsjön began to become turbid from algal blooms caused by increased nutrient supply. The phytoplankton biomass increased and, due to the heavy growth of blue-green algae and the large amounts of coliform bacteria, bathing was periodically prohibited. At that time, as a young girl, the first author swam regularly in Western Basin. In August, after swimming, she and her brothers and sisters broke out in skin rashes, associated with blue-green algal blooms. She observed small, green balls, about 2 mm in diameter, probably consisting of colonies of *Gloeotrichia echinulata* (Cronberg, 1999). In 1968, the first known algal poisoning in Sweden occurred near the shore of Eastern Basin of Lake Ringsjön. The death of cows was assumed to be caused by their consuming of toxic blue-green algae. The hooves, venter and muzzle of the dead cows examined, had a blue tone due to the blue-green algae (Cronberg, 1983). At the time of the cattle death, the dominant algae was *Anabaena flos-aquae*.

Study site

Lake Ringsjön is situated in southernmost Sweden (55° 52′ N; 13° 32′ O) and consists of three basins: Sätofta Basin (4.2 km^2, mean depth 3.0 m, maximum depth 17.0 m), Eastern Basin (20.5 km^2, mean depth 6.1 m, maximum depth 16.4 m) and Western Basin (14.8 km^2, mean depth 3.1 m, maximum depth 5.4 m)(Hansson et al., 1999). Like many other lakes in urban and agricultural areas, Lake Ringsjön became eutrophicated during the 1960s and 1970s. In the beginning of the 1980s, a *nutrient reduction programme* was initiated, which reduced the external input of nutrients to the lake from phosphorus loadings exceeding 30 tons per year, to around 10 tons per year. About ten years later (1989–1990), a *cyprinid fish reduction programme* (biomanipulation) was carried out in Sätofta Basin removing about 100 tons of fish (about 60% of the standing stock). During the winter of 1988, there was a fish-kill in Eastern Basin, eliminating about 500 tons of fish (about 80%) (Hamrin, 1999). The biomanipulation in Western Basin, which was performed in 1992, removed about 120 tons of fish (about 50%).

The fish reduction programme performed during 1988 and 1992 resulted in increased water transparency and reduced concentration of nutrients in the water phase (Enell et al., 1996). After these positive changes in the ecosystem of Eastern Basin, the present study aimed to examine the occurrence of toxic algae following the reduction in nutrients and fish.

Methods

Sampling

On 3 August and 2 September 1994, algal scum was sampled with 45 μm plankton nets near the shore. In 1995, at the end of June, before visible blooms of blue-green algae occurred, weekly sampling was started for the analysis of phytoplankton and the blue-green algal toxin microcystin. The sampling was terminated at the end of October, when the blooms had disappeared.

Table 1. Potentially toxic blue-green algae recorded in Lake Ringsjön samples

Aphanizomenon flos-aquae BRÉB.
A. klebahnii (ELENK.) PECH. & KALINA.
Anabaena circinalis RABENH.
A. flos-aquae BRÉB. ex BORN. et FLAH.
Microcystis aeruginosa KÜTZ.
M. botrys TEIL.
M. flos-aquae (WITTR.) KIRCHN.
M. robusta (CLARK) NYG.
M. viridis (A. BR) LEMM.
M. wesenbergii KOM. in KONDR.
Woronichinia naegeliana (UNG.) ELENK.

Using a Plexiglass tube, water was sampled from the surface to a depth of 2 m. The water was collected in a plastic bucket. This integrated sample was used for the analysis of phytoplankton and microcystins. Phytoplankton samples for quantitative analysis were preserved with Lugols solution. The samples for microcystin analysis were kept cool during transport to the laboratory. Phytoplankton samples for qualitative analysis were collected with a 25 μm plankton net and preserved in 2–4% formalin. Dominant algal species recorded are listed in Table 1. As soon as blue-green algal blooms were noticeable, algal scum for toxicity tests was sampled near the shore by using a plankton net with a mesh size of 45 μm. The samples were kept cool during the transport to the laboratory where they were deep-frozen prior to lyophilization. The lyophilized algae were kept frozen until the toxicity test could be performed.

Analyses

The qualitative investigation of phytoplankton was done using a normal light microscope. The species were recorded and their relative frequency estimated. The quantitative analysis of phytoplankton was carried out with an inverted phase contrast microscope. The samples settled in 5 ml chambers for at least 12 hours. The dominant taxa were counted and their biomass was calculated according to Cronberg (1982). For HPLC analysis of microcystins, 4 l of the quantitative samples were filtered through GF/C filters. The filters were air-dried and kept frozen until the HPLC preparation could be performed. The HPLC analysis was carried out according to Lawton et al. (1994). The toxicity of the phytoplankton community was tested by mouse bioassay, according to Skulberg et al. (1994). The observed symptoms include three categories:

1. neurotoxic: rapid death within 30 minutes after injection
2. hepatotoxic: death occurring 1–4 hours after injection, resulting in a blood-filled liver and liver necrosis
3. protracted toxic: (death occurring 4–24 hours after injection without any observable signs of organ damage.

Results

Algal toxicity

In September 1981, the first mouse bioassay was carried out . The toxicity test showed that the blue-green algal bloom, consisting of *Microcystis* spp., was hepatotoxic. In July and August 1985–1987, algal blooms in all three basins of Lake Ringsjön were tested by mouse bioassays. On most occasions, the blooms were hepatotoxic and were dominated by *Microcystis viridis* and *M. wesenbergii* (Table 1). In 1987, *Anabaena flos-aquae* and *A. circinalis* were also frequently occurring. In 1981, 1985, 1986 and 1987, a total of 13 samples of algal scum from the three basins of the lake were tested for toxicity by mouse bioassay (Table 2). Nine of the 13 samples were hepatotoxic, while the remaining four were non-toxic.

In 1994, two samples of scum taken on August 3rd and September 2nd, were both hepatotoxic according to the mouse bioassays. Microcystins were not detected in the August sample by HPLC, while 1 071 $\mu g\ g^{-1}$ dry weight was detected in the September sample. In 1995, all 11 scum samples collected from July 12th–October 24th were hepatotoxic according to the mouse bioassay. The time between injection and death varied from 43 to 121 minutes (Table 3). Microcystin concentrations varied between 0 and 742 ng l^{-1} (Figure 1).

Algal composition

In 1994, the algal composition differed in the two samples; the August sample was dominated by the blue-green algae *Microcystis wesenbergii* and *M. aeruginosa*, while *M. flos-aquae* and *Aphanizomenon klebahnii* dominated the September sample. In the spring of 1995, the phytoplankton in Eastern Basin was dominated by diatoms, mainly species belonging to the genera *Aulacoseira, Cyclotella* and *Stephanodiscus*. In June, the diatoms disappeared almost completely, and the blue-green algae increased in abundance. From mid-June–September, the total algal biomass consisted from 58 to 93% of blue-greens.

Microcystis was most often the dominant genus. In 1995, *Microcystis wesenbergii* was generally the most common species and also reached the highest biomass of all algal species in Eastern Basin. In October, however, *Aphanizomenon klebahnii* dominated (Table 4). In August, blue-green algae showed a maximum at 6 mg l^{-1}.

Discussion

The microcystin concentration (1 071 $\mu g\ g^{-1}$ dry weight) in the sample of hepatotoxic scum from Eastern Basin in September 1994 is the highest reported in Swedish lakes so far. The maximum concentration of microcystins found in algae from the nearby Lakes Vombsjön and Finjasjön were 274 and 152 $\mu g\ g^{-1}$ dry weight, respectively (Annadotter, 1993). This hepatotoxic sample from Eastern Basin was dominated by *Microcystis flos-aquae* and *Aphanizomenon klebahnii*. Neither of these species have yet been reported as being microcystin-producers even though *A. klebahnii* has been shown to produce an unidentified, protracted toxin (Cronberg & Annadotter, 1996). It is thus likely that the microcystin in the September sample was produced by *Microcystis flos-aquae* or *A. klebahnii*. The genus *Aphanizomenon* is generally known as a neurotoxin producer, while several *Microcystis* taxa, such as *M. aeruginosa* and *M. viridis*, are common producers of microcystins.

During the period June–October 1995, several microcystin maxima were recorded in Eastern Basin. The highest maximum was noted on 12 July (742 ng l^{-1}; Table 4). On that occasion, *Microcystis botrys* was dominant. In addition, the blue-green algae *Microcystis wesenbergii* and *M. aeruginosa* were abundant. The algal biomass was, however, not as high in July as in August–September. During this later period, the microcystin maxima were much lower (Figure 1). Thus the blue-green algal toxins were not related with high blue-green algal biomass. Contradictory results have been reported by Kotak et al. (1995) who found a positive correlation between the concentration of

Table 2. Dominant blue-green algal species and results from mouse bioassays in Lake Ringsjön, 1981–1987. Sources: 1. Mattsson & Willén (1986); 2. Mattsson & Willén (1987a,b); 3. Cronberg (1983)

	Mouse bioassay	Dominant bluegreen algal species	Sources
Sätofta Basin			
26 Jul. 1985	hepatotoxic	*Microcystis viridis* & *M. wesenbergii*	1
28 Aug. 1985	hepatotoxic	*Microcystis viridis* & *M. wesenbergii*	1
24 Jul. 1986	hepatotoxic	*Microcystis viridis* & *M. wesenbergii*	2
22 Jul. 1987	non toxic	*Anabaena circinalis, A. flos-aquae, M. wesenbergii, M. aeruginosa*	2
Eastern Basin			
Sep. 1968	not tested	*Anabaena flos-aquae*	3
16 Sep. 1981	hepatotoxic	*Microcystis* spp.	3
25 Jul. 1985	hepatotoxic	*Microcystis viridis*	1
26 Aug. 1985	hepatotoxic	*Microcystis viridis* & *M. wesenbergii*	1
2 Jul. 1986	hepatotoxic	*Microcystis viridis, M. wesenbergii* & *M. aeruginosa*	2
22 Jul. 1987	non-toxic	*M. wesenbergii, M. viridis* & *M. aeruginosa*	2
Western Basin			
26 Jul. 1985	hepatotoxic	*Microcystis viridis, M. wesenbergii* & *Anabaena* spp.	1
28 Aug. 1985	hepatotoxic	*Microcystis viridis, M. wesenbergii* & *Aphanizomenon flos-aqaue*	1
22 Jul. 1987	non-toxic	*A. circinalis, Aphanizomenon flos-aquae, M. wesenbergii, M. aeruginosa*	2

Table 3. Dominant blue-green algal species and the results of mouse bioassays in Eastern Basin of Lake Ringsjön from June to October 1995

Date	Mouse Bioassay death time (min)	Dominant blue-green algal species (net samples) (blue-green algal taxa in dominance order)
29-Jun-95	not tested	*Microcystis wesenbergii, Microcystis botrys, Woronichinia naegeliana*
05-Jul-95	not tested	*M. wesenbergii, Microcystis flos-aquae, M. botrys*
12-Jul-95	hepatotoxic (84)	*M. wesenbergii, W. naegeliana, M. flos-aquae, M. botrys*
19-Jul-95	not tested	*M. flos-aquae, W. naegeliana, M. botrys, Microcystis aeruginosa*
26-Jul-95	hepatotoxic (65)	*M. wesenbergii, W. naegeliana, M. flos-aquae, M. botrys, Microcystis viridis*
02-Aug-95	hepatotoxic (60)	*W. naegeliana, M.wesenbergii, M. botrys, M. aeruginosa, Microcystis robusta*
09-Aug-95	hepatotoxic (100)	*M. wesenbergii, W. naegeliana, M. flos-aquae, M. botrys, M. viridis*
16-Aug-95	hepatotoxic (82)	*M. wesenbergii, W. naegeliana, M. flos-aquae, M. aeruginosa, M.botrys*
23-Aug-95	hepatotoxic (82)	*M. wesenbergii, W. naegeliana, M. flos-aquae, Aphanizomenon klebahnii, M. botrys*
30-Aug-95	not tested	*M. wesenbergii, W. naegeliana, M. viridis*
06-Sep-95	hepatotoxic (99)	*A. klebahnii, M. wesenbergii, W. naegeliana*
13-Sep-95	hepatotoxic (121)	*M. wesenbergii, M. viridis, A. klebahnii, W. naegeliana, M. botrys, M. flos-aquae*
22-Sep-95	hepatotoxic (99)	*M. wesenbergii, M. viridis, W. naegeliana, Aphanizomenon klebahnii*
27-Sep-95	not tested	*M. wesenbergii, M. botrys*
03-Oct-95	not tested	*M. wesenbergii, A. klebahnii, W. naegeliana, M. viridis*
16-Oct-95	hepatotoxic (100)	*W. naegeliana, Anabaena* sp.
24-Oct-95	hepatotoxic (43)	*W. naegeliana, Anabaena flos-aquae, M. wesenbergii, M. botrys, A. klebahnii*

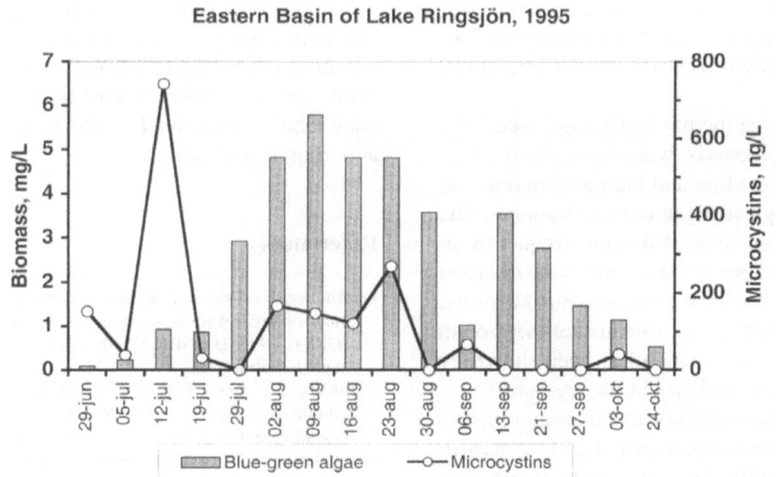

Figure 1. Blue-green algal biomass (mg l^{-1}) and concentrations of microcystin (ng l^{-1}) in Eastern Basin of Lake Ringsjön from June to October 1995.

Table 4. Dominant blue-green algal species, biomass and microcystin concentrations in the Eastern Basin of Lake Ringsjön from June to October 1995

Date	HPLC analyses Microcystins,	Biomass (mg l^{-1})	Dominant blue-green algae (quantitative samples) (Percentage of total blue-green algal biomass given within brackets)
29-Jun-95	153	0.09	*Microcystis wesenbergii* (40%), *Woronichinia naegeliana* (23%)
05-Jul-95	39	0.22	*M. wesenbergii* (43%), *Microcystis flos-aquae* (29%), *Microcystis botrys* (13%)
12-Jul-95	742	0.92	*M. botrys* (43%), *M. wesenbergii* (38%)
19-Jul-95	32	0.86	*M. wesenbergii* (42%), *M. flos-aquae* (26%), *W. naegeliana* (15%)
26-Jul-95	0	2.92	*M. wesenbergii* (32%), *M. flos-aquae* (26%), *W. naegeliana* (15%), *Microcystis robusta* (10%)
02-Aug-95	167	4.82	*W. naegeliana* (44%), *M. wesenbergii* (19%), *M. robusta* (9%)
09-Aug-95	148	5.78	*M. wesenbergii* (40%), *M. flos-aquae* (21%), *W. naegeliana* (18%)
16-Aug-95	122	4.81	*M. wesenbergii* (36%), *M. botrys* (22%), *M. flos-aquae* (15%), *W. naegeliana* (10%)
23-Aug-95	269	4.81	*M. wesenbergii* (44%), *M. viridis* (19%), *W. naegeliana* (18%), *Aphanizomenon klebahnii* (12%)
30-Aug-95	0	3.57	*M. wesenbergii* (58%), *M. viridis* (24%)
06-Sep-95	66	1.02	*A. klebahnii* (38%), *M. wesenbergii* (26%), *M. viridis* (19%)
13-Sep-95	0	3.54	*M. wesenbergii* (61%), *M. viridis* (24%), *A. klebahnii* (11%)
22-Sep-95	0	2.76	*M. wesenbergii* (47%), *M. viridis* (17%), *A. klebahnii* (17%)
27-Sep-95	0	1.45	*M. botrys* (30%), *M. wesenbergii* (26%), *M. viridis* (20%)
03-Oct-95	41	1.14	*A. klebahnii* (51%), *M. wesenbergii* (23%), *M. viridis* (15%)
24-Oct-95	0	0.52	*A. klebahnii* (25%), *Anabaena flos-aquae* (18%), *M. wesenbergii* (23%), *M. flos-aquae* (15%)

microcystin-LR and of chlorophyll *a*, while Wicks & Thiel (1990) found a weak, negative correlation between the total microcystin content and chlorophyll *a* concentration.

On four occasions (26 July; 13 and 22 September, and 24 October), the scum samples were hepatotoxic, while microcystins were not detected in the quantitative analysis of the water samples (Tables 2 & 4). The discrepancy could be explained by the possibility of the hepatotoxin detected by mouse bioassay not being a microcystin. On the other hand, the scum samples were collected near the shore and concentrated by a plankton net, while the unconcentrated water samples were collected in the pelagial as an integrated sample from the surface to a depth of 2 m. In other words, the net sample and the water sample were collected on the same occasion, but not at the exact same spot. It is well known that algal toxicity can vary between sites located only a few metres apart (Gorham & Carmichael, 1979).

Despite the fact that the phytoplankton, especially the blue-green algal biomass, was drastically reduced following nutrient reduction and biomanipulation, the toxicity of the blue-green algae did not diminish. The microcystin-producing algae still occur frequently in Lake Ringsjön. It appears that there is no relation between the trophic state of the lake and algal toxicity. Oligotrophic water bodies can also sustain the growth of toxic blue-green algae, which has been shown by Mez et al. (1997) in alpine lakes in Switzerland.

Despite the nutrient reduction and biomanipulation of Lake Ringsjön, toxic blue-green algal blooms still appeared in 1994–1995. In fact, the toxin production in 1995 was at its highest in July (742 ng l^{-1}), while the algal biomass was low (1 mg l^{-1}). At the biomass maximum in August (6 mg l^{-1}; Figure 1), the microcystin concentration was only 148 ng l^{-1}. Since several toxigenic species co-occurred, the toxin production could not be correlated with any specific blue-green algal species.

The study from 1995 showed that hepatotoxins were present in Eastern Basin from the end of June until the end of October. This continuous toxicity during blue-green algal blooms constitutes a permanent health risk for swimmers and animals that drink the water. Eastern Basin drains into Western Basin, where the water-works, Ringsjöverket, is situated. Until 1987, Western Basin was used as a raw water supply for the municipalities of Landskrona, Kävlinge and Eslöv, among others. In 1987, the source of raw water was changed. Drinking water was drawn from Lake Bolmen, southern Småland instead, via an underground tunnel. However, in July 1995, the tunnel collapsed and Western Basin of Lake Ringsjön was again used as the raw water reservoir. As a part of an investigation of toxic algae in South Swedish Lakes, algae and water were collected several times also from Western Basin. Samples from surface water of this Basin on 6 September 1995 contained 722 ng microcystins l^{-1}, and a concentrated algae sample taken on the same day was hepatotoxic according to a mouse bioassay. The sample was dominated by *Microcystis wesenbergii, M. robusta* and *M. botrys* (Cronberg & Annadotter, 1996). Hence, toxic blue-green algae are not restricted to the Eastern Basin, but also occur in the basin used as a water reservoir.

Major conclusions from this study are that both *Microcystis flos-aquae* and *Aphanizomenon klebahnii* are potential producers of microcystin, which has never been reported before. Moreover, we found no quantitative relation between the biomass of blue-green algae and toxin production.

References

Annadotter, H., 1993. Cyanobacterial toxins in drinking water – a study of two Swedish water plants and a literature study. VA-FORSK 1993-03. ISBN 91-88392-42-2. (In Swedish with an English abstract).

Bergman, E., 1997. Ringsjön. Restaurering genom cyprinidreduktion – effekten av fiskreduktionen. Rapport. Limnologiska avdelningen, Lunds Universitet. (In Swedish).

Carmichael, W. W., 1992. Cyanobacteria secondary metabolites – the cyanotoxins. J. appl. Bacteriol. 72: 445–459.

Cohen, S. G. & C. B. Reif, 1953. Cutaneous sensitization to blue-green algae. J. Allergy 24: 452–457.

Cronberg, G., 1982. Phytoplankton changes in Lake Trummen induced by restoration. Fol. Limnol. Scand. 18: 1–119.

Cronberg, G., 1983. Ringsjöområdet – Ekosystem i förändring. Report. Institute of Limnology, University of Lund. ISSN 0348–0798. (In Swedish).

Cronberg, G., 1999. Qualitative and quantitative investigations of phytoplankton in Lake Ringsjön. Hydrobiologia 404: 27–40.

Cronberg, G. & H. Annadotter, 1996. Förekomst av algtoxiner i sjöar inom Malmöhus län 1994–1995. Report. Institute of Ecology, University of Lund. (In Swedish).

Dillenberg, H. O. & M. K. Dehnel, 1960. Toxic waterbloom in Saskatchewan, 1959. Can. Med. Assoc. 83: 1151–1154.

Enell, M., J. Fejes & C. Williams, 1996. Ringsjön – ekosystem i förändring 1995. Report. Mellanskånes Kommunalförbund. (In Swedish).

Falconer, I. R., A. M. Beresford & M. T. Runneger, 1983. Evidence of liver damage by toxin from a bloom of the blue-green alga *Microcystis aeruginosa*. Med. J. Aust. 28: 511–514.

Francis, G., 1878. Poisonous Australian lake. Nature 18: 11.

Gorham, P. R. & W. W. Carmichael, 1979. Phycotoxins from blue-green algae. Pure Appl. Chem. 52: 165–174.

Hamrin, S., 1999. Planning and execution of the fish reduction in lake Ringsjön. Hydrobiologia 404: 59–63.

Hansson, L.-A., M. Enell & E. Bergman, 1999. Lake Ringsjön: its catchment area, its history and its importance. Hydrobiologia 404: 1–7.

Hawkins, P. R., M. T. Runnegar, A. R. Jackson & I. R. Falconer, 1985. Severe hepatotoxicity caused by the tropical cyanobacterium (blue-green alga) *Cylindrospermopsis raciborskii* (Woloszynska) Seenaya and Subba Raju isolated from a domestic water supply reservoir. Apl. Envir. Microbiol. 50: 1292–1295.

Heise, H. A., 1949. Symptoms of hay fever caused by algae. J. Allergy 20: 383–385.

Heise, H. A., 1951. Symptoms of hay fever caused by algae. II. *Microcystis*, another form of algae producing allergenic reactions. Ann. Allergy 9: 100–101.

Kotak, B. G., A. K.-Y. Lam, E. E. Prepas, S. L. Kenefick & S. E. Hrudey, 1995. Variability of the hepatotoxin microcystin-LR in hypertrophic drinking water lakes. J. Phycol. 31: 248–263.

Lawton, L. A., C. Edwards & G. A. Codd, 1994. Extraction and High-performance Liquid Chromatographic Method for the Determination of Microcystins in Raw and Treated Waters. Analyst 119: 1525–1530.

Lundberg, R., L. Edler, S. Fernö, M. G. Lind & P. O. Nilsson, 1983. Algförgiftning hos hund. Svensk veterinärtidning 35: 509–516.

Main, C. D., P. H. Berry, R. L. Peet & J. P. Robertson, 1977. Sheep mortalitites associated with the blue green algae *Nodularia spumigena*. Australian Veterinary Journal 53: 578–581.

Mattsson, R. & T. Willén, 1986. Toxinbildande blågröna alger i svenska insjöar 1985. Swedish Environmental Protection Agency. ISBN 91-620-3096-5. (In Swedish with an English abstract).

Mattsson, R. & T. Willén, 1987a. Vattenblommande och toxinbildande alger. Report from the Veterinary National Institute, Uppsala, Sweden. (In Swedish).

Mattsson, R. & T. Willén, 1987b. Vattenblomning av blågrönalger och deras toxicitet. Analysresultat 1987. Report from the Veterinary National Institute, Uppsala, Sweden. (In Swedish).

Mez, K., K. A. Beattie, G. A. Codd, K. Hanselmann, B. Hauser, H. Naegeli & H. H. R. Preisig, 1997. Identification of microcystin in benthic cyanobacteria linked to cattle deaths on alpine pastures in Switzerland. Eur. J. Phycol. 32: 111–117.

Pouria, S., A. Andrade, J. Barosa, R. L. Cavalcanti, V. T. S. Barreto, C. J. Ward, W. Preiser, G. K. Poon, G. H. Neild & G. A. Codd, 1998. Fatal microcystin intoxication in haemodialysis unit in Caruaru, Brazil. The Lancet 352: 21–26.

Pybus, M. J. & D. P. Hobson, 1986. Mass mortality of bats due to probable blue-green algal toxicity. J. Wild. Dis. 22: 449–450.

Rose, E. T., 1953. Toxic algae in Iowa lakes. Proc. Iowa Acad. Sci. 60: 738–745.

Skulberg, O., 1998. Blågrønnalger, giftprodusenter i vann. Populaervitenskapelig Magasin 6: 52–55.

Skulberg, O., B. Underdal & H. Utkilen, 1994. Toxic waterblooms with cyanophytes in Norway – current knowledge. Algological Studies 75: 279–289.

Schwimmer, M. & D. Schwimmer, 1964. Algae and medicine. In D. J. Jackson (ed.), Algae and Man. Plenum Press, New York.

Wicks, R. J. & P. G. Thiel, 1990. Environmental factors affecting the production of peptide toxins in floating scums of the Cyanobacterium *Microcystis aeruginosa* in a hypertrophic African reservoir. Env. Sci. Technol. 24: 1413–1418.

Hydrobiologia **404:** 131–144, 1999.
L.-A. Hansson & E. Bergman (eds), Nutrient Reduction and Biomanipulation as Tools to Improve Water Quality: The Lake Ringsjön Story.

Simulating the effects of biomanipulation on the food web of Lake Ringsjön

Anders Persson[1], Andreas Barkman[2] & Lars-Anders Hansson[1]
[1]*Department of Limnology, Insitute of Ecology, Ecology Building, University of Lund, SE-223 62 Lund, Sweden*
[2]*Chemical Engineering II, Center for Chemistry and Chemical Engineering, P.O. Box 124, SE-221 00 Lund, Sweden*

Key words: phosphorus, model, excretion, grazing

Abstract

A dynamic, process-oriented, deterministic and phosphorus-based model was developed to simulate the food web dynamics of Lake Ringsjön, in particular the long-term effects of biomanipulation in terms of reduction of omnivorous fish. The model contains 14 state variables, each with a differential equation describing sources and sinks of phosphorus. The state variables encompass piscivorous and omnivorous fish, zooplankton, phytoplankton, sediment and lake water. The model simulates densities of fish and phytoplankton adequately, both before and after biomanipulation, although the actual lake phytoplankton density varied more year-to-year compared to the model predictions. According to the model, a biomanipulation will cause an increase in zooplankton biomass. This prediction contradicts available field data from the lake which do not indicate any significant change in zooplankton biomass resulting from the performed biomanipulation. This discrepancy may partly be attributed to structural uncertainties in the model, related to the size structure of predators on zooplankton, i.e. the omnivorous fish community. The simulations suggest that phosphorus was routed along the pelagic food chain to a larger extent after omnivorous fish were removed, whereas the amount of phosphorus routed via the sediment and benthivorous fish decreased following fish removal. Accordingly, translocation of phosphorus from sediment to water by benthivorous fish is predicted to be substantially reduced by biomanipulation, resulting in an overall reduction in the release of new phosphorus to phytoplankton. Irrespective of simulated fishing effort (reduction of $\leq$0.5% d^{-1} for two years), the model predicts that P-release from the sediment and the external load will remain sufficiently high to force the system back to its previous state within a decade. Thus, recurrent biomanipulations and/or combined abatement strategies may be necessary to maintain low phytoplankton density. Known structural model uncertainties may however affect the robustness of such detailed predictions about the system resilience.

Introduction

Eutrophication is a world wide problem in lakes exploited by man, often resulting in extensive blooms of blue-green algae. These blooms cause severe environmental problems including toxic drinking water and low oxygen levels; accordingly, there is a need for tools that can predict critical amounts of nutrient load when blooms are likely to occur, and can predict the outcome of various restoration techniques. Model simulations of lake ecosystems is one such tool that has been used. In the 1960s, phosphorus loading and phytoplankton biomass were found to be correlated and phosphorus was found to be the limiting nutrient for phytoplankton production in most lakes (Sakamoto, 1966). Simple regression models where developed to predict phytoplankton biomass at a certain nutrient loading (Vollenweider, 1976). Later, eutrophication models incorporating both nutrient loading and zooplankton grazing were developed to account for some of the unexplained variability in the early regression models. Today, general view states that the potential primary productivity is set by nutrient loading, but that interactions within the lake food web determines the resulting phytoplankton biomass (Carpenter et al., 1985).

When lakes are eutrophicated, not only the quantity, but also the quality of different food web components change. Phytoplankton are favored before submerged vegetation in the competition for light and nutrients (Wetzel & Hough, 1973), resulting in a shift towards a more homogenous lake habitat. Piscivorous fish suffer when the visibility of the water decreases and the habitats preferred by their young, in terms of macrophyte beds, disappear (Diehl, 1988; Persson et al., 1988). Omnivorous fish increase in biomass as predation pressure decreases and zooplankton prey becomes a limited resource, favoring fish species that are able to feed on small individuals and alternative resources such as benthic invertebrates and detritus (Lessmark, 1983; Persson, 1983). Thus, the food web may change from a system controlled by piscivorous fish, with a zooplankton community consisting of large and efficient grazers that buffer changes in phytoplankton production, to a system dominated by omnivorous fish keeping zooplankton density and grazing at low levels. Reducing the external load of nutrients is not always enough to bring the system back to one controlled by piscivorous fish (Moss et al., 1996). Theoretical studies have suggested that each of these states represents an equilibrium stabilized by a number of feedback mechanisms (Scheffer, 1991).

Biomanipulation has been suggested as a method to restore the food web of eutrophicated lakes, forcing the system from one equilibrium to another (Shapiro et al., 1975). The most common aim of biomanipulation has been to increase grazing loss on the phytoplankton community by decreasing fish predation on zooplankton. Indeed, reduction of planktivorous fish biomass often does produce clear water conditions initially, but the probability of sustained effects depends on the nutrient loading and the development of stabilizing feedback mechanisms such as occurrence of large *Daphnia* (Lammens et al., 1990). Experiences from the field have also shown that changes in fish community composition could have profound effects on phytoplankton productivity by a number of different indirect effects and subtle interactions which affect nutrient availability (Vanni, 1996). Many fish species use resources from different habitats and trophic levels which may promote or suppress phytoplankton productivity depending on the net flux of elements (Brabrand et al., 1990; Schindler et al., 1996). When lakes of northern Europe are eutrophicated, the fish community typically shifts from percids to cyprinids (Persson et al., 1988, 1991). Once eutrophic, the lake often has a high density of omnivorous fish such as roach and bream. By feeding on the sediment, these species potentially act as a pump, translocating sediment phosphorus to the water column (Lamarra, 1975; Andersson et al., 1978; Persson, 1997b).

A number of recent studies imply that fish excretion could be a driving force in the P-cycle (Carpenter et al., 1992; Persson, 1997a; Vanni & Layne, 1997, Vanni et al., 1997). In an oligotrophic system, dual migrations by fish increased productivity by translocating phosphorus from littoral to pelagic habitats (Schindler et al., 1993). Vanni & Layne (1997) concluded that food web effects on nutrient recycling, especially excretion by consumers, were important in controlling phytoplankton productivity and dynamics. Thus, both experimental and field evidence state convincingly that consumer recycling of nutrients is an important process in the trophic cascade and that it should be included in models of food web dynamics (Persson, 1997a; Vanni & Layne, 1997).

The objective of this study is to apply an already developed mathematical model to investigate anticipated long-term effects of biomanipulation on phytoplankton biomass in Lake Ringsjön. Moreover, the aim is to use the model in evaluating the potential role of phosphorus translocated and recycled within the water column by consumers, and to assess the effects of biomanipulation on the excretion by consumers.

Material and methods

Study site

Lake Ringsjön is situated in southernmost Sweden (55° 52′ N; 13° 32′ O) and consists of three basins: Sätofta Basin (4.2 km^2, mean depth 3.0 m, maximum depth 17.0 m), Eastern Basin (20.5 km^2, mean depth 6.1 m, maximum depth 16.4 m) and Western Basin (14.8 km^2, mean depth 3.1 m, maximum depth 5.4 m) (Hansson et al., 1999). As with many other lakes in urban and agricultural areas, Lake Ringsjön became eutrophicated during the 1960s and 1970s. In the beginning of the 1980s, a *nutrient reduction programme* was initiated, which reduced the external input of nutrients to the lake from phosphorus loadings exceeding 30 tons per year to around 10 tons per year. About ten years later (1989–1990), a *cyprinid fish reduction programme* (biomanipulation) was carried out in Sätofta Basin removing about 100 tons of fish (about 60% of the standing stock). During the winter of 1988, there was a fish-kill in Eastern Basin, eliminating about 500 tons of fish (about 80%) (Hamrin, 1999). The biomanipulation in Western Basin, which was performed in 1992, removed about 120 tons of fish (about 50%).

Model description

The utilized food web model is dynamic, process-oriented and deterministic. The model is phosphorus based and it is developed specifically to address long-term food web dynamics of Southern Swedish lakes, and particularly those lakes subjected to eutrophication. Consequently, the model was developed to simulate and to evaluate long-term dynamics in terms of annual and summer average biomasses, rather than detailed abundances and seasonal dynamics of various species. The applied model approximates the lake to a continuously mixed tank reactor. Consequently, it is spatially isotropic with respect to all its state variables. This application of the model is constrained to Sätofta Basin and Eastern Basin, since these two basins form a separate system, and Western Basin constitutes the outflow of Eastern Basin. The model consists of 14 state variables representing phosphorous stored in fish, zooplankton, phytoplankton, water and in the sediment. The conceptual structure of the model is displayed in Figure 1. For the phytoplankton and zooplankton, the model structure resembles that previously described in Scavia et al. (1988). Since the model focuses on long-term dynamics of the food web, structural components with long-term intrinsic properties have been given special attention. Consequently, the fish populations in the model have relatively low structural unreliability with respect to processes and feedback mechanisms. The equations and parameters which govern the fluxes of phosphorus between the different pools are described in more detail in Persson & Barkman (1997).

Forcing functions

Inflow of phosphorus for the period 1977–1992 was based on measured values of flow and phosphorus concentrations in the water of the tributaries (Figure 2a). The seasonal pattern was set constant using 1992 as a model year. For the period before 1977, few data are available between which interpolations were made, and for the period following 1992, the 1992 condition was used.

State variables

The dynamics of each state variable is described by a differential equation using a mass balance approach of sources and sinks of phosphorus. For organism components, a major source is the consumption/uptake of phosphorus, r_c (ton P yr^{-1}), which is modelled according to

$$r_c = O \cdot k \cdot \mathrm{FR} \cdot f(T) \cdot f(l), \qquad (1)$$

where O is the density of the organism (ton P) and k is the instantaneous rate of increase (yr^{-1}). FR denotes the functional response and is described using Michaelis–Menten kinetics:

$$\mathrm{FR} = \frac{[N_{\mathrm{eff}}]^n}{[N_{\mathrm{eff}}]^n + K^n}, \qquad (2)$$

where $[N_{\mathrm{eff}}]$ denotes the efficient concentration/density of resources and K the half saturation constant (concentration/density units). *f(T)* adjusts the uptake/consumption rate as a function of temperature, and *f(l)* describes phosphorus uptake by phytoplankton and consumption rate of fish as a function of light intensity. Piscivores, in the model perch *(Perca fluviatilis)*, have been shown to be more sensitive to low light conditions than planktivores, in the model roach *(Rutilus rutilus)* (Diehl, 1988), and consequently this was accounted for by $f(l)$. The efficient concentration of resources, $[N_{\mathrm{eff}}]$, was modelled according to

$$[N_{\mathrm{eff}}] = s_1 \cdot N_1 + s_2 \cdot N_2 + \cdots + s_i \cdot N_i, \qquad (3)$$

where N_i represents different prey types and s_i the selectivity coefficient for each prey type. Here, *Daphnia* was considered to be a more efficient consumer of blue-green algae than copepods and small grazers (Burns, 1968). Likewise, small piscivores were considered more efficient consumers of copepods than planktivores due to escaping behavior of copepods, and the opposite was considered for fish feeding on small grazers due to their small size (Lessmark, 1983).

The constant n in Equation 2 was set to 1 for phytoplankton, zooplankton and piscivorous fish, resulting in a type II functional response. A type III functional response (n=2) was considered to be more appropriate for fish feeding on zooplankton, since the species in Lake Ringsjön often have a mixed diet also including benthic prey (Persson & Hansson, 1999). This would result in a lower consumption rate of zooplankton at low prey densities, which is often modelled as a sigmoid functional response. Benthic feeding was described as:

$$r_c = P \cdot k \cdot (1 - \mathrm{FR}) \cdot f(T), \qquad (4)$$

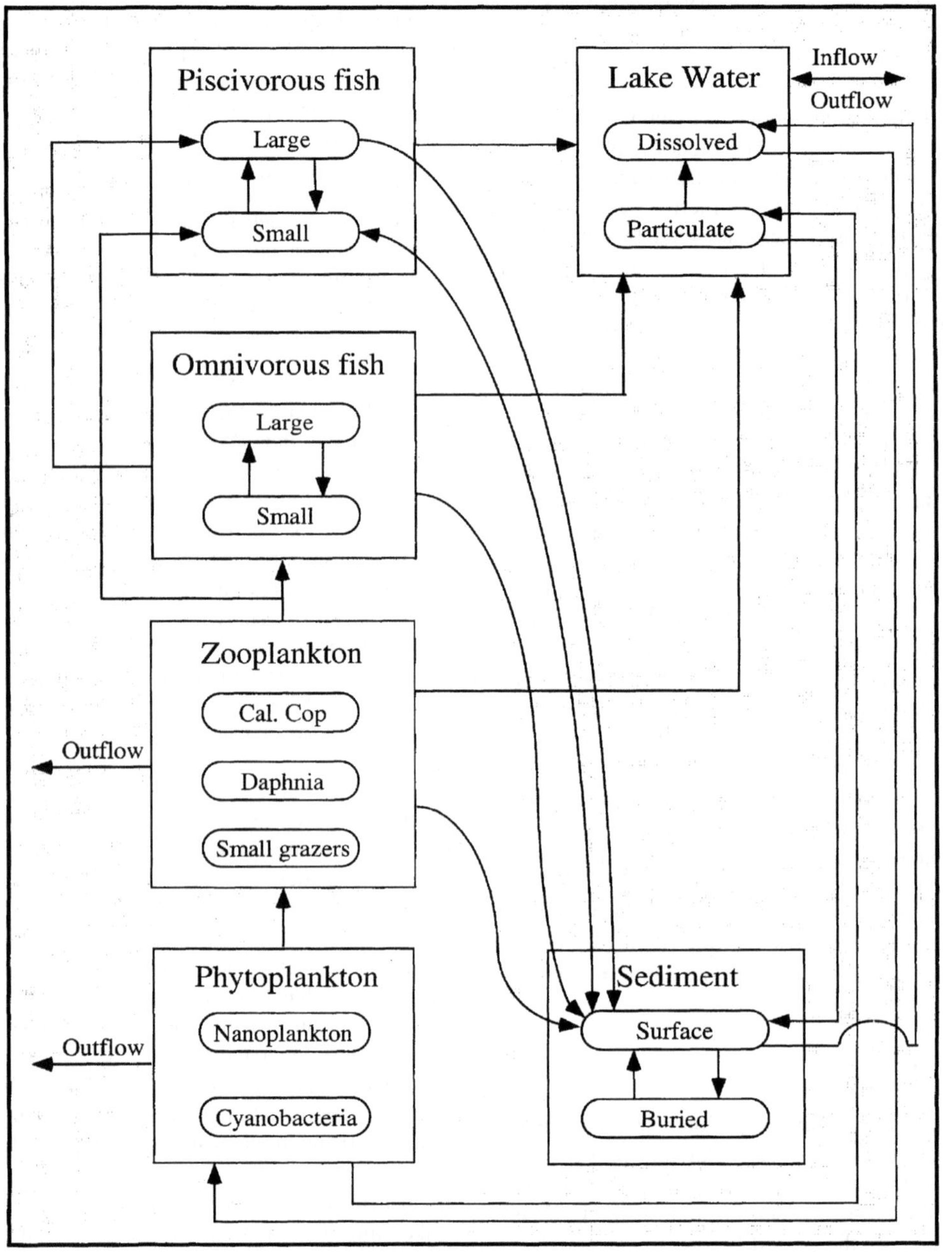

Figure 1. A conceptual description of the lake food web model, with six major phosphorus compartments, including: water, sediment, phytoplankton, zooplankton, omnivorous and piscivorous fish. The lake water compartment is divided into dissolved and particulate phosphorus and the sediment into surface and buried phosphorus. Each organism compartment is divided into two or three size-classes.

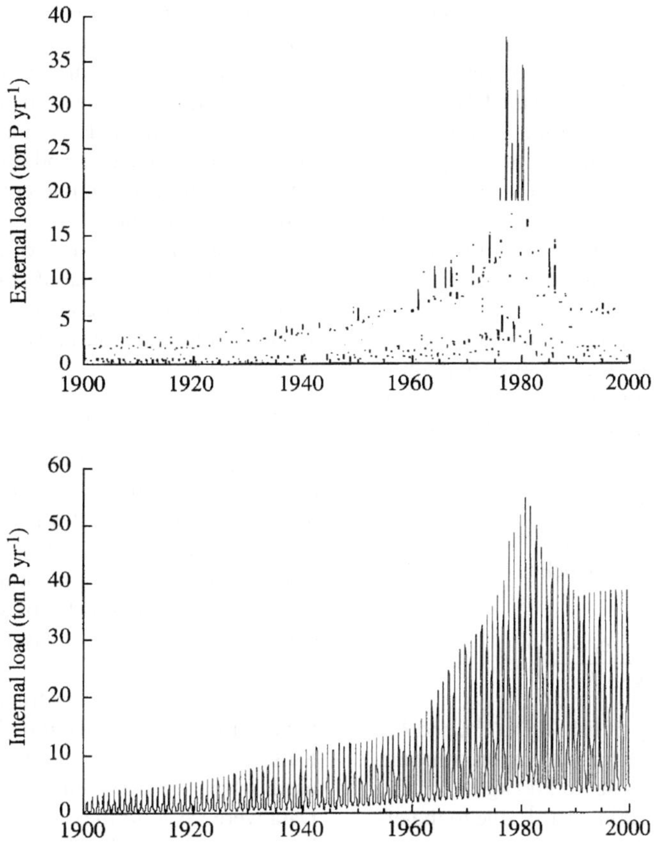

Figure 2. External loading of phosphorus and simulated release rate of phosphorus from the sediment of Lake Ringsjön at a 'no fishing' scenario between year 1900 and 2000.

where FR is the functional response in the Equation (2) describing the consumption of their main prey. For benthic prey the instantaneous intake rate, k, was set lower than for their main diet, i.e. benthic feeding was considered an alternative strategy when zooplankton prey were scarce (Lessmark, 1983; Persson, 1983).

Of the consumed phosphorus, some is egested, some is allocated to growth, and yet some is excreted. The net growth rate, r_G (ton P yr^{-1}) is modelled according to

$$r_{G_i} = r_{C_i} - r_{EX_i} - (1 - d_{\text{eff}}) \cdot r_{C_i}, \quad (5)$$

where the digestive efficiency of the ith component, d_{eff}, was set constant. We used the relationships from Brabrand et al. (1990) based on field estimates for phosphorus excretion by benthic feeding roach and perch. For zooplanktivorous and piscivorous feeding, we used relationships of growth efficiency obtained for roach and perch in laboratory by Lessmark (1983). For zooplankton, the excretion rate, r_{EX} (ton P yr^{-1}), was assumed to be proportional to the respiration rate and modelled as dependent on food availability according to Scavia et al. (1988):

$$r_{EX_i} = r_{RC1_i} + FR_i \cdot r_{RC2_i}, \quad (6)$$

where FR represents the type II functional response in Equation (2). In order to reflect the higher mass specific recycling rate of small zooplankton compared to *Daphnia*, small zooplankton were given higher values of r_{RC1} and r_{RC2} than *Daphnia*.

Apart from grazing, phytoplankton losses are due to natural mortality within the water column and sedimentation. In the model, we did not distinguish between these two processes, and instead a temperature dependent loss rate, r_L (ton P yr^{-1}), based on the sinking rate of phytoplankton, was used according to

$$r_{L_i} = P_i \cdot \frac{k_m}{z_m} \cdot f(T), \quad (7)$$

Blue-green algae were modelled to have a lower loss rate (k_m = 15) than nanoplanktonic algae (k_m = 50) since many species have gas vacuoles to prevent sedimentation (Reynolds, 1984). The studied lakes are shallow and wind induced resuspension is an important process which keeps large amounts of phosphorus within the water column. In order to mirror longer residence times the flux of dead algae was modelled as an input to a pool of particulate phosphorus, which was subsequently subjected to both remineralization and sedimentation.

For zooplankton, the natural mortality rate was set to zero because fish predation was assumed to dominate zooplankton losses. The natural mortality, r_m(ton P yr^{-1}), of small fish was modelled as a function of food availability according to

$$r_{m_i} = P_i \cdot k_{m_i} \cdot (1 - \mathrm{FR}_i), \tag{8}$$

where k_m is the natural mortality rate constants (yr^{-1}).

Mortality of large fish was modelled to be density dependent according to

$$r_{m_i} = P_i \cdot \left(k_1 + k_2 \cdot \frac{\rho_i}{\rho_i + K_{M_i}}\right), \tag{9}$$

where ρ_i denotes the density of fish (kg ha^{-1}), and K_m represents a natural mortality half saturation constant. For omnivorous fish this was set to 300 kg ha^{-1}, and for piscivorous fish lower, 100 kg ha^{-1}, due to the effect of cannibalism within the piscivorous community (Craig, 1996). To reflect the average length of life the constants k_1 and k_2 were set to 0.1 and 0.3 for both piscivorous and omnivorous fish.

Dead fish, zooplankton and particulate phosphorus were transported to the sediment where it was mineralized. Diffusion was assumed to be the rate controlling step for P-flux from sediment to the water. The mass transfer of phosphorus from the sediment to the water was modelled via a mass transfer coefficient, expressed as a ratio between the molecular diffusion coefficient and the thickness of the diffusionary boundary layer, as well as by the concentration gradient between interstitial water and lake water. The interstitial concentration of dissolved phosphorus was assumed to be controlled by the equilibrium between dissolved and adsorbed phosphorus. Accordingly, the interstitial concentration in the sediment was modelled in proportion to the total amount of phosphorus in the upper sediment. This ratio was calibrated to achieve approximate measured interstitial phosphorus concentrations during both oligotrophic and eutrophic conditions. Surface sediments in eutrophic lakes often show higher phosphorus concentrations than buried sediment, and we therefore divided the sediment in two sections: surface sediment and buried sediment. In the surface sediment, the exchange of phosphorus with lake water is facilitated by organisms (Gallepp, 1979). This was modelled as a random process by increasing the molecular diffusion rate compared to the physical diffusion coefficient.

Fish recruitment was expressed as the transference of a fraction biomass/ phosphorus of large fish to small fish during spawning, thereby contributing to a new cohort once a year. The transfer was modelled according to

$$r_{\mathrm{TRANS}i} = P_i \cdot k_{\mathrm{TRANS}i} \cdot h_{\mathrm{eff}}, \tag{10}$$

where k_{TRANS} was set to 0.1 for piscivorous fish and 0.08 for omnivorous fish. h_{eff} denotes the hatching efficiency of fish eggs and was set to 0.8 for both types of fish. In the model, piscivorous fish was assumed to spawn in April, and omnivorous fish in May. Aging was modelled as a transfer of small fish biomass at a constant rate of 0.6 yr^{-1}, assuming an average age of small fish to 1.6 years.

The model was constructed using MATLAB/ SIMULINK and was integrated with a 4-step Runge–Kutta integration method. A variable time step was used, where the maximum step was set to 0.005 year. Data were saved to files every 0.02 year (i.e. approximately once a week). These data files were used to produce figures and to do further calculations (i.e. summer averages and annual averages).

Simulated scenarios

The model was used to simulate different scenarios of biomanipulation in terms of selective reduction of omnivorous fish. The simulations covered biomanipulation efforts from no reduction of omnivorous fish to a maximum effort of 0.5% reduction of the biomass per day. This range covers the efforts made in Lake Ringsjön which has been estimated to approximately 0.3% d^{-1} (see 'Result' section below). For all scenarios, the fish reduction started in 1988 and lasted for two years. The model was validated using data on phytoplankton biomass (wet weight; 1975–1995), zooplankton (dry-weight; 1987–1995) and fish (wet weight; 1987 and 1990). Biomass of phytoplankton were only available for 1987–1994, so chlorophyll *a* estimates for the period 1975–1986 were converted to phytoplankton biomass using a regression model based on data for chlorophyll *a* and phytoplankton biomass between

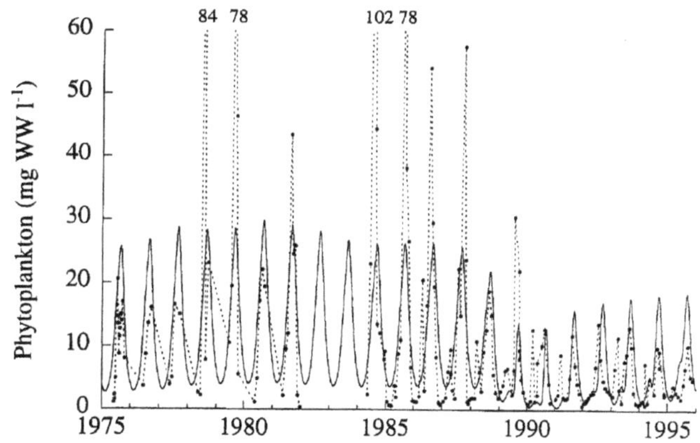

Figure 3. Simulated (line) and measured (dotted line) biomass of phytoplankton in Lake Ringsjön 1975–1995.

1987 and 1994. Rotifers were only counted during the years 1990–1993; since rotifers could account for a substantial fraction of the total zooplankton biomass (5–40% of total zooplankton biomass in Lake Ringsjön 1990–1993), we corrected for that in the estimates lacking rotifer counts. For the years 1990–1993, the average monthly percentage rotifer biomass of total zooplankton biomass was calculated; this percentage was then used to correct the biomass estimates of the years 1987–1989 and 1994.

Results

A comprehensive description of the model characteristics during the eutrophication phase is described in more detail elsewhere (Persson & Barkman, 1997). Consequently, only a brief description is presented followed by results of different biomanipulation scenarios. With the onset of eutrophication (Figure 2), zooplankton biomass responds rapidly, leading to variable algae biomass ('paradox of enrichment') that suffers from a high grazing pressure. However, this period is relatively short in the simulation ($\approx$20 years), and is replaced by another period where phytoplankton and omnivorous fish respond to enrichment in a rather similar way. At the peak of external load, piscivorous fish suffer from impaired light conditions, and are consequently decreasing in importance as predators, leaving the field open to omnivorous fish to expand in biomass in response to increased resource availability. Even if the external load is lowered drastically, the sediment release of P holds the system in a highly productive phase (Figure 2). In paradox, lowering the external load results in a decreased biomass of zooplankton because the resource base (phytoplankton) responds faster than the predators (omnivorous fish).

Only two estimates of fish density were available; before and after the fish reduction. The model estimates of the planktivore fish density before fish reduction is in agreement with field estimates (Figure 4). Running the model with a fish reduction effort representing approximately 0.3% d^{-1} for two years removed 622 tons of fish, which approximates the estimated removal in the lake of 600 tons. Also, when using this fishing effort in the model, the predicted fish density resembles the fish density estimated in the field (Figure 4).

The simulated zooplankton biomass was generally higher than measured values (Figure 4). The model predicts an increase in zooplankton as well as a decrease in phytoplankton density as a result of the biomanipulation (Figure 4). This prediction was in agreement with observed phytoplankton densities (Figure 3), but was not fully supported by field data for zooplankton densities. Data available on phytoplankton biomass 1975–1995 display a substantial year to year variation from peaks up to 100 mg l^{-1} some years to low values below 5 mg l^{-1} (Figure 3). As expected from the modelling objectives and current model structure, the model is not able to simulate this year to year variability. Nevertheless, it manages to adequately reproduce phytoplankton densities representing average values during this period of time. Furthermore, the effects of fish reduction on phytoplankton biomass is predicted adequately, with the exception of occasional peaks (Figure 3).

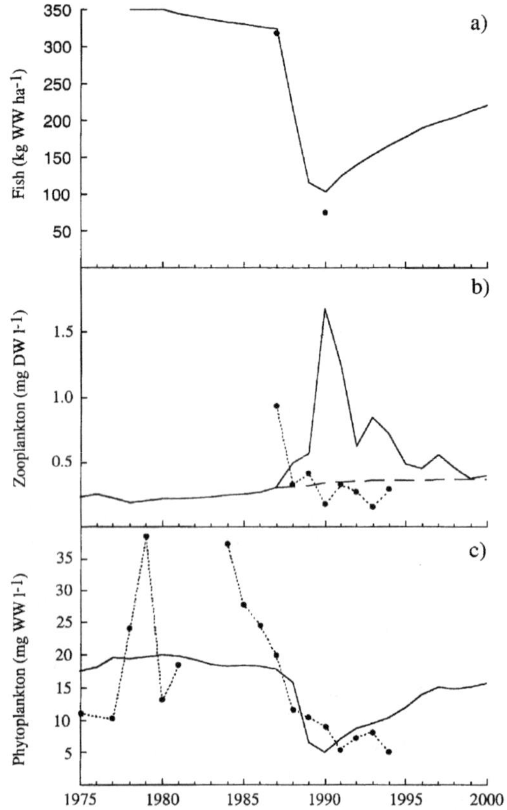

Figure 4. Simulated density of fish (a), zooplankton (b), and phytoplankton (c) (line) at a fishing effort of 0.3% d^{-1}. Dots indicate measured biomasses in Lake Ringsjön 1975–1995. Fish biomass is given as annual average, and phyto- and zooplankton as summer average (May–September). Also given is simulated zooplankton density at a 'no fishing' scenario (broken line).

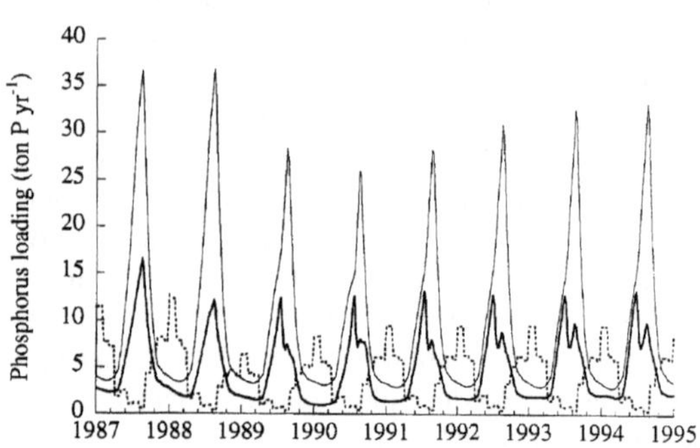

Figure 5. Estimated external load of phosphorus (broken line), simulated sediment release (thin line) and excretion by consumer (thick line) in Lake Ringsjön. The simulations were performed with a fish reduction between 1988 and 1990 at a rate of 0.3% d^{-1}.

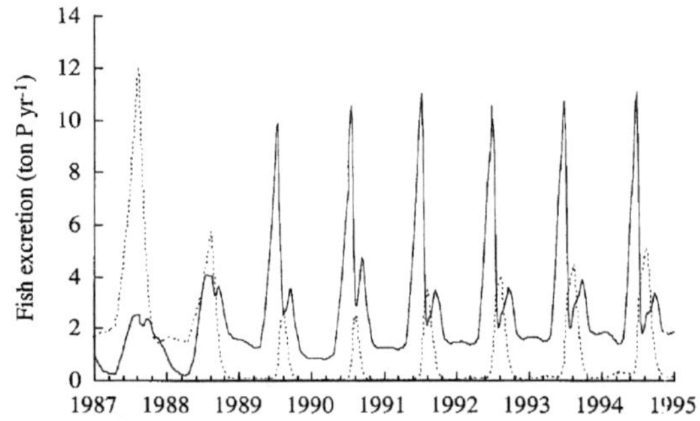

Figure 6. Simulated excretion of phosphorus from fish consuming pelagic prey (broken line) and benthic prey (solid line). In the simulation, omnivorous fish are removed between 1988 and 1990 at a rate of 0.3% d^{-1}.

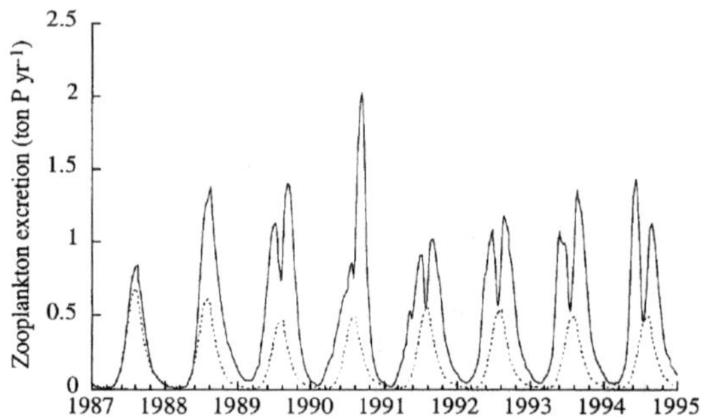

Figure 7. Simulated excretion by zooplankton at a 'no fishing' (broken line) and fishing scenario. In the simulation, omnivorous fish are removed between 1988 and 1990 at a rate of 0.3% d^{-1}.

Phosphorus release from the sediment constitutes the dominating portion of new phosphorus released to phytoplankton growth in the model. However, both external loading and excretion by consumers constitute major sources (Figure 5). In the model, most phosphorus excreted by consumers before the biomanipulation is released by omnivorous fish, mostly due to feeding on benthic resources (Figure 6). The model results indicate that piscivorous fish have an excretion rate comparable to zooplankton excretion. This can only be explained by the zooplankton resource being of less importance for the transport of phosphorus to higher trophic levels compared to benthic resources. The simulations indicate that excretion by consumers is reduced as a result of the biomanipulation (Figure 5). This is particularly true for the transport of phosphorus from the sediment to the water by benthic feeding fish, which is substantially reduced by the

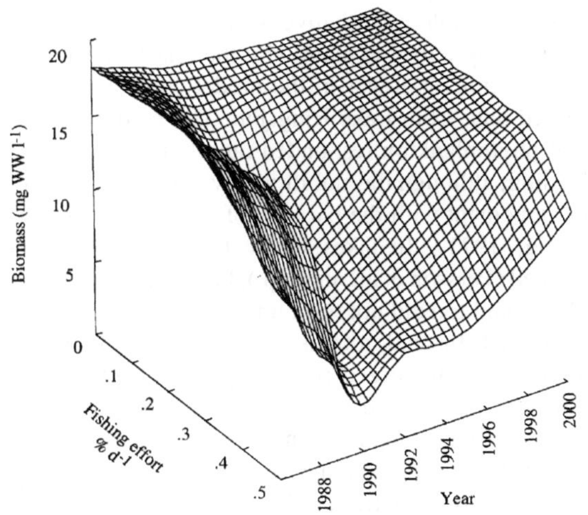

Figure 8. Simulated biomass of phytoplankton. The fish reduction was simulated to be of a constant efficiency (0–0.5% d^{-1}), lasting for two years between 1988 and 1990. The figures shows that at high fishing intensity, the response lasts longer than at lower fishing intensity.

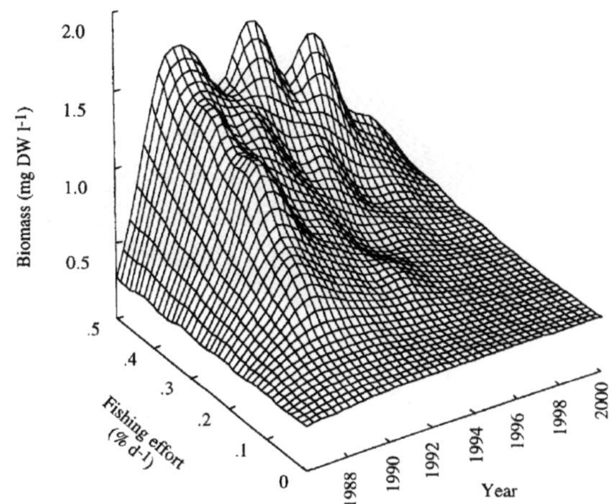

Figure 9. Simulated biomass of zooplankton. The fish reduction was simulated to be of a constant efficiency (0–0.5% d^{-1}), lasting for two years between 1988 and 1990.

biomanipulation, while the recycling within the water column is increased (Figs 6 and 7).

Investigation of different simulated biomanipulation scenarios, from no fishing to a high fishing rate (0 – 0.5% d^{-1}), provides an indication of the anticipated minimum and maximum effects of biomanipulation. The simulations predict that piscivorous fish increase only when a relatively high fishing rate is applied (Figure 11), i.e. when light conditions are improved enough to ensure that small piscivores are relieved from competition from planktivores. However, the piscivore population eventually decreases to former densities within about 10 years. Omnivorous fish is initially reduced by up to 90% (0.5% d^{-1}), but will gradually increase to reach former densities within 30 years (Figure 10). Zooplankton is predicted to become substantially affected by fish removal at all fishing rates (Figure 9). Available field data however do not support these predictions (Figure 4). The effects of the biomanipulation on zooplankton biomass is predicted to gradually decrease, and former densities are reached within 15 years after fishing. The effects of fish reduction are predicted to cascade down to the phytoplankton community, although effects of a low fishing effort are anticipated to be minor and short-term (Figure 8). When simulating a high fishing rate, phytoplankton is initially reduced to very low (< 5 mg l^{-1}) abundances, but the phytoplankton community

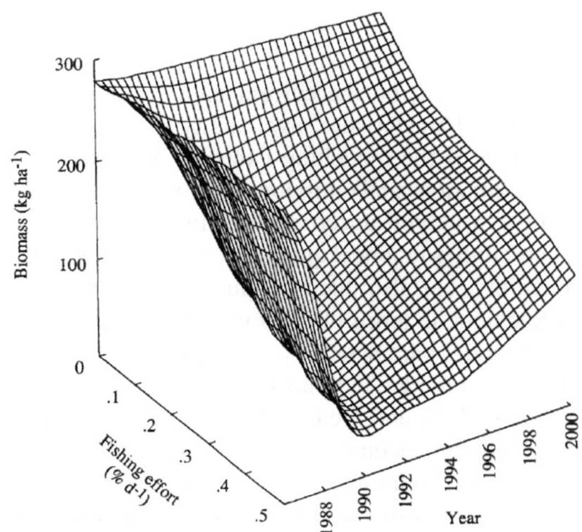

Figure 10. Simulated biomass of omnivorous fish. The fish reduction was simulated to be of a constant efficiency (0–0.5% d^{-1}), lasting for two years between 1988 and 1990.

eventually recovers, reaching former densities after approximately 10 – 15 years (Figure 8).

Discussion

A model is a system describing another system. As such, models always involve simplifications and generalizations of the 'true' or suspected characteristics of the modelled system. Simplifications are natur-

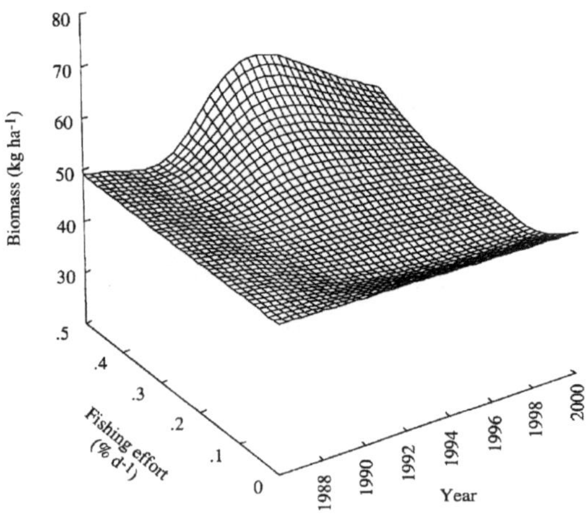

Figure 11. Simulated biomass of piscivorous fish. The fish reduction was simulated to be of a constant efficiency (0–0.5% d^{-1}), lasting for two years between 1988 and 1990.

ally dependent on the current understanding of the system's characteristics, availability of data, and the objective of the modelling exercise. Simplifications, approximations and omissions of processes inevitably involve structural uncertainty, or model unreliability. Uncertainties in model quantities are due to model unreliability, but also due to technical uncertainties, or inexactness, in terms of random and systematic errors in all model parameters. An increased structural reliability of a model, in terms of an increased number of state-variables and feed-back mechanisms, increases the need for adequate data but often also the error propagation within the model, and therefore the inexactness of the calculated output. Hence, all modelling exercises involve trade-offs between unreliability of the model structure and inexactness of the calculated output. Trade-offs are made in order to achieve an acceptable quality of the model and the calculated results in relation to the overall objective of the study. In this sense, model uncertainty is regarded as an attribute of the information produced, whereas model quality is a pragmatic relation between a model and its intended usage/users (Funtowicz & Ravetz, 1992).

In this modelling context, the trade-offs made have naturally involved exclusion of some processes and inclusion of others. In the following, two processes causing model unreliability are discussed in more detail as they are likely to have important consequences for the interpretation of the results.

The macrophytes were omitted in the model due to lack of reliable quantitative data. This model unreliability may partly explain why the resistance against perturbations is lower in the model compared to observed data from Lake Ringsjön. The model predicts that phytoplankton responded rapidly to increased P-loading, although the stability may have been affected in agreement with the paradox of enrichment (Rosenzweig, 1971). In lakes where the model has been applied (Persson et al., in prep.), field data suggest that phytoplankton did not respond as rapidly as the model suggested. Instead, a transition phase of an increased amount of submerged macrophytes took place. Not until the submerged macrophytes crashed did blue-green algae become the dominant producers on a year to year basis, which is in agreement with observations both in field (Scheffer et al., 1997) and experimental conditions (Jasser, 1995). The lake studied here (i.e. Western Basin excluded) has a mean depth of 5.6 m, and 26% of the lake area is below 3 m depth, suggesting that submerged macrophytes could colonize large areas, thereby competing with phytoplankton for nutrients, and also providing habitats for young piscivorous fish and large grazers (Benndorf, 1990). This factor may have contributed to the delayed food web response in the lake compared to model predictions. In Lake Ringsjön, however, macrophytes did not exhibit any significant response to biomanipulation (Strand, unpubl.). Hence, any discrepancy between model prediction and field data for the period concerning the biomanipulation, is not likely to be attributed to the exclusion of macrophyte interactions in the model. Despite this, we acknowledge that the stability of the model response to biomanipulation may be different should macrophytes be included in the model. This may particularly be the case if applying the model to shallow lakes where interactions between algae and macrophytes are important (van Donk et al., 1990).

Another process formulation, which is likely to introduce considerable model unreliability is the sediment–water interaction. As has been mentioned previously, the P-flux from the sediment to the water is modelled as a diffusion process, i.e. diffusion is assumed to be the rate-controlling step for the P-flux. The advantage of the adopted approach, adjustment of the interstitial P-concentration as a function of the total amount of P in the upper sediment, is that the model could easily be applied to different lakes by changing the parameters governing the P-concentration. However, the lumped approach does not address factors such as pH in the overlaying water and the redox potential of the sediment, and their influence on the interstitial concentration and the P-flux. Quantitative relationships between these factors and P-release from sediments in Lake Ringsjön have been established in laboratory essays. These investigations are however inconsistent with field observations (Boström, 1982). Due to this, and foreseen increased structural complexity, the above mentioned factors have been excluded from the model. An obvious shortcoming with the adopted approach is that the model probably underestimates the release of phosphorus from the sediment during hypertrophic conditions. Hence, short-term but significant releases of P from the sediment due to these factors will therefore remain unadressed.

The model suggests that the biomass of zooplankton increases in response to fish reduction, suggesting biomanipulation to be a useful method of restoring lakes. As mentioned above, available field data from Lake Ringsjön show that changes in grazing did not occur although large amounts of omnivorous fish were removed. One explanation of this discrepancy between modelled and observed data may be the structure of the modelled omnivorous fish population. In the model, omnivorous fish is divided into two different classes (see Figure 1). This model design reduces the possibility of addressing the effects of a shift in size-structure as a result of biomanipulation. A shift in the fish community towards smaller individuals may compensate for reduced biomass in terms of consumption rate of zooplankton. Indeed, mean fish weight decreased after fish removal and numerical abundance increased in Lake Ringsjön (Bergman et al., this volume). The simulation predicts a larger fraction of small omnivorous fish compared to large omnivorous fish. Thus, the model addresses, within its structural limits, the change in fish community composition caused by the fish reduction in Lake Ringsjön. As a result, zooplankton consumption is predicted to increase and consumption of benthic resources decrease. This change in diet however is due both to a shift towards smaller omnivorous fish and, contrary to field observations – an increased zooplankton biomass.

The effect of fish activity on phytoplankton is twofold; both indirectly by determining grazer abundance and directly by releasing nutrients. When Lake Ringsjön was eutrophicated, the fish community shifted from dominance of perch and high ratio between piscivorous and omnivorous fish, to dominance of roach and bream and low piscivorous to omnivorous fish ratio. Before fish reduction, benthivorous bream constituted more than half of the total fish biomass (Bergman et al., 1999). Since fish reduction was most efficient in removing bream, the importance of bream as a benthic-pelagic link of P is therefore likely to have been reduced.

The results of the simulations suggest that a fish reduction may have large effect on the rate of P-release to phytoplankton, although the total release of P from consumers may only decrease marginally. Also, the P-release from fish consuming benthic resources, i.e. fish translocating phosphorus from the sediment to the water, is predicted to be substantially reduced. Prior to fish reduction, fish biomass in the model was to a large extent built up by resources of benthic origin, as indicated by the excretion rate for piscivorous fish being comparable to zooplankton excretion. Thus, the transfer of P was not only directed along the pelagic food chain (phytoplankton–zooplankton-planktivorous fish–piscivorous fish), but to a larger extent routed via sedimented phytoplankton (blue-green algae not grazed by zooplankton) providing, directly or indirectly via benthic invertebrates, a resource pool for benthivorous fishes. After fish reduction, when zooplankton densities in the model increased, P from benthic feeding fishes was supplied to the water at a much lower rate and the route of P along the pelagic food chain increased in importance. Thus, the simulations indicate that trophic structure may affect not only the rate at which P is supplied for phytoplankton growth, but also the routes by which P is transported from phytoplankton to higher trophic levels; either via detrivores decomposing dead phytoplankton or via zooplankton grazing on living phytoplankton. Similar conclusions have also been made for other eutrophic lakes, where changes in the abundance of omnivorous/benthivorous fish cascade down to phytoplankton via changes in nutrient availability rather than grazing pressure from zooplankton (Horppila, 1994, Blindow et al. in prep.).

A critical aspect when using biomanipulation as a restoration technique for eutrophicated lakes is whether or not the resulting conditions will remain stable (Shapiro, 1990). The model predicts that a low fishing effort will only have short-term effects and that the system will rapidly return to its previous unfavorable state. As an example, removing 525 tons of fish (0.2% d^{-1}) is predicted to decrease phytoplankton biomass for approximately five years. Removing an additional 150 tons (0.4% d^{-1}) is predicted to have effects on phytoplankton biomass that will last for almost twice that time. Thus, a low fishing effort would only have short-term effects, leaving little opportunity for feedback mechanisms, such as colonization of submerged macrophytes, to come into action. A large fishing effort, however, is predicted to have effects lasting for more than 10 years, resulting in opportunities for positive feedback mechanisms to be active for a longer period of time, thereby yielding in a higher probability of a successful biomanipulation. In this sense, the model predictions are validated by data from the lake, the fish reduction having a low efficiency since fish escaped by moving between the different basins (Bergman et al., 1999). This has, in Lake Ringsjön, resulted in relatively marginal effects on phytoplankton biomass and a tendency to a return to previous conditions (Bergman et al., 1999). Field experiences from other biomanipulation experiments support the results from the simulations indicating that the cascading effects are positively correlated with fishing effort (Hansson et al., 1998).

Regardless of fishing effort used in the simulations, the model predicts that the lake will return to its previous state. It may therefore be necessary to perform biomanipulation repeatedly in order to keep phytoplankton biomass in Lake Ringsjön at low levels. Due to the above mentioned structural uncertainties, which may affect stability of the system, the frequency of repeated fish reduction may differ from the ones suggested by the simulations. According to the model, the unfavorable trend is probably caused by the relatively marginal reduction of the total P-load following the biomanipulation (see Figure 5). Consequently, combined abatement strategies, including reduction of the total P-load, may be needed in order to permanently mitigate the situation, or to reduce the frequency of repeated biomanipulations. However, extensive reductions of P from external sources have already been implemented in the Lake Ringsjön catchment area (M. Enell, unpubl.). Further reductions of the external load were therefore regarded as unrealistic, and not included in the current scenario analysis. We do however, acknowledge that joint implementation of measures, directed towards reducing the total P-load, could alter, or off-set, the necessary frequency of biomanipulation.

Conclusions

The applied model predicts that the ecosystem of Lake Ringsjön could be altered from a turbid to a clear water state by reduction of omnivorous fish, and that the response, in terms of phytoplankton biomass. is correlated with the fishing intensity.

The model predicts that a fish reduction will affect phytoplankton abundance in two ways; by increasing zooplankton grazing, and by decreasing the excretion of new P by benthic feeding fish.

The model is adequately validated for both phytoplankton and fish, but not for zooplankton biomass. This suggests that the phytoplankton abundance is not primarily affected by the fish community through cascading effects along the pelagic food chain, as expected from food chain theory. Thus, the fish community may have other significant effects on the nutrient dynamics in Lake Ringsjön than the ones included in the model. Re-parameterization of the applied model is not likely to offset this discrepancy between data and model prediction, unless uncertainties in model structure are mitigated. However, increasing structural complexity is hazardous because suggested processes to be included are not known quantitatively.

The model indicates that the system will return to its previous unfavorable state, regarding phytoplankton biomass, within 5–20 years depending on biomanipulation intensity. The model also indicates that a major factor for this is total load of P, keeping the potential productivity at a high level. Thus, from a management perspective, repeated biomanipulations, or combined abatement strategies, may be necessary.

Acknowledgments

We thank Prof. Harald Sverdrup, Department of Chemical Engineering II, Lund University, and Dr Stellan Hamrin, Institute of Freshwater Research, for initiating the modelling exercise by developing the first embryo of the present model. Field data for validation was provided by Dr Eva Bergman and the Swedish Environmental Research Institute (Dr Magnus Enell). This study was supported by grants from National Swedish Environmental Protection Agency (AP, AB), the Municipality of Hässleholm (AB), and Swedish Natural Science Research Council (L.-A.H).

References

Andersson, G., H. Berggren, G. Cronberg & C. Gelin, 1978. Effects of planktivorous and benthivorous fish on organisms and water chemistry in eutrophic lakes. Hydrobiologia 59: 9–15.

Benndorf, J., 1990. Conditions for effective biomanipulation: Conclusions derived from whole-lake experiments in Europe. Hydrobiologia 200/201: 187–203.

Bergman, E., S. F. Hamrin & P. Romare, 1999. The effects of cyprinid reduction on the fish community. Developments in Hydrobiology. Hydrobiologia 404: 65–75.

Boström, B., 1982. Recycling of nutrients from lake sediment. Ph.D. thesis, University of Uppsala, Sweden.

Brabrand, Å., B. A. Faafeng & J. P. M. Nilssen, 1990. Relative importance of phosphorus supply to phytoplankton production: fish excretion versus external loading. Can. J. Fish. Aquat. Sci. 47: 364–372.

Burns, C. W., 1968. The relationship between body size of filter-feeding Cladocera and the maximum size particle ingested. Limnol. Oceanogr. 13:675–678.

Carpenter, S. R., J. F. Kitchell & J. R. Hodgson, 1985. Cascading trophic interactions and lake productivity. BioScience 35: 634–639.

Carpenter, S. R., C. E. Kraft, R. Wright, X. He, P. A. Soranno & J. R. Hodgson, 1992. Resilience and resistance of a lake phosphorus cycle before and after food web manipulation. Am. Nat. 140: 781–798.

Craig, J. F., 1996. Pike. Biology and exploitation. Fish and Fisheries series 19. Chapman & Hall, London, U.K.

Diehl, S., 1988. Foraging efficiency of three freshwater fishes: effects of structural complexity and light. Oikos 53: 207–214.

Funtowicz, S. O. & J. R. Ravetz, 1992. The emergence of post-normal science. In R. Schomberg (ed.), Science, Politics and Morality. Scientific Uncertainty and Decision Making. Kluwer Academic Publishers: 85–123.

Gallepp, G. W., 1979. Chironomid influence on phosphorus release in sediment-water microcosms. Ecology 60: 547–556.

Hamrin, S., 1999. Planning and execution of the fish reduction in Lake Ringsjön. Hydrobiologia 404: 59–63.

Hansson, L.-A., H. Annadotter, E. Bergman, S. F. Hamrin, E. Jeppesen, T. Kairesalo, E. Luokkanen, P-Å. Nilsson & Ma. Søndergaard, 1998. Biomanipulation as an application of food chain theory: constraints, synthesis and recommendations for temperate lakes. Ecosystems 1: 558–574.

Hansson, L.-A., M. Enell & E. Bergman, 1999. Lake Ringsjön: its catchment area, its history and its importance. Hydrobiologia 404: 1–7.

Horppila, J., 1994. Interactions between roach (*Rutilus rutilus* (L.)) stock and water quality in Lake Vesijärvi (southern Finland). Ph.D. thesis, University of Helsinki, Finland.

Jasser, I., 1995. The influence of macrophytes on a phytoplankton community in experimental conditions. Hydrobiologia 306: 21–32.

Lamarra, V. A. Jr., 1975. Digestive activities of carp as a major contributor to the nutrient loading of lakes. Verh. Int. Ver. Limnol. 19: 2461–2468.

Lammens, E. H. R. R., R. D. Gulati, M.-L. Meijer & E. van Donk, 1990. The first biomanipulation conference: a synthesis. Hydrobiologia 200/201: 617–627.

Lessmark, O., 1983. Competition between perch (*Perca fluviatilis*) and roach (*Rutilus rutilus*) in south Swedish lakes. Ph.D. thesis, Inst. of Limnology, Lund Univ., Lund, Sweden. 172 pp.

Moss, B., J. Stansfield, K. Irvine, M. G. Perrow & G. Phillips, 1996. Progressive restoration of a shallow lake: a 12-year experiment in isolation, sediment removal and biomanipulation. J. Appl. Ecology. 33: 71–86.

Persson, A., 1997a. Effects of fish predation and excretion on the configuration of aquatic food webs. Oikos 79: 137–146.

Persson, A., 1997b. Phosphorus release by fish in relation to external and internal load. Limnol. Oceanogr. 42: 577–583.

Persson, A. & A. Barkman, 1997. Modelling lake food web dynamics. In A. Persson, Consumption Patterns and Excretion in Aquatic Food Webs. Ph.D. thesis. Dept. of Ecology, Limnology, Lund University, Sweden.

Persson, A. & L.-A. Hansson, 1999. Diet shift in fish following competitive release. Can. J. Fish. Aquat. Sci. 56: 70–78.

Persson, L., 1983. Food consumption and the significance of detritus and algae to intraspecific competition in roach (*Rutilus rutilus*) in a shallow eutrophic lake. Oikos 41: 118–125.

Persson, L., G. Andersson, S. F. Hamrin & L. Johansson, 1988. Predator regulation and primary production along the productivity gradient of temperate lake ecosystems. In S. R. Carpenter (ed.), Complex Interactions in Lake Communities. Springer Verlag, New York, NY: 45–65.

Persson, L., S. Diehl, L. Johansson, G. Andersson & S. F. Hamrin, 1991. Shift in fish communities along the productivity gradient of temperate lakes-patterns and the importance of size-structured interactions. J. Fish Biol. 38: 281–293.

Reynolds, C. S., 1984. The ecology of freshwater phytoplankton. Cambridge University Press, Cambridge, UK.

Rosenzweig, M. L., 1971. Paradox of enrichment: destabilization of exploitation ecosystems in ecological time. Science 171: 385–387.

Sakamoto, M., 1966. Primary production by phytoplankton community in some Japanese lakes and its dependence on lake depth. Arch. Hydrobiol. 62: 1–28.

Scavia, D., G. A. Lang & J. F. Kitchell, 1988. Dynamics of Lake Michigan plankton: a model evaluation of nutrient loading, competition, and predation. Can. J. Fish. Aquat. Sci. 45: 165–177.

Scheffer, M., 1991. Fish and nutrient interplay determines algal biomass: a minimal model. Oikos 62: 271–282.

Scheffer, M., S. Rinaldi, A. Gragnani, L. R. Mur & E. H. van Nes, 1997. On the dominance of filamentous cyanobacteria in shallow, turbid lakes. Ecology 78: 272–282.

Schindler, D. E., J. F. Kitchell, X. He, S. R. Carpenter, J. R. Hodgson & K. L. Cottingham, 1993. Food web structure and phosphorus cycling in lakes. Trans. Am. Fish. Soc. 122: 756–772.

Schindler, D. E., J. F. Kitchell, X. He, S. R. Carpenter, J. R. Hodgson & K. L. Cottingham, 1996. Food web structure and littoral zone coupling to pelagic trophic cascades. In G. A. Polis & K. O. Winemiller (eds), Food Webs: Integration of Patterns and Dynamics. Chapman & Hall, New York, NY: 96–105.

Shapiro, J., 1990. Biomanipulation: the next phase – making it stable. Hydrobiologia 200/201: 13–27.

Shapiro, J., V. Lamarra & M. Lynch, 1975. Biomanipulation: an ecosystem approach to lake restoration. In P. L. Brezonic & J. L. Fox (eds), Proc. Symp. on Water Quality Management Through Biological Control. University of Florida: 85–96.

Strand, J., 1999. The development of submerged macrophytes in Lake Ringsjön after biomanipulation. Hydrobiologia 404: 113–121.

van Donk, E., M. P. Grimm, R. D. Gulati & J. P. G. Breteler, 1990. Whole-lake food-web manipulation as a means to study community interactions in a small ecosystem. Hydrobiologia 200/201: 275–289.

Vanni, M. J., 1996. Nutrient transport and recycling by consumers in lake food webs: implications for algal communities. In G. A. Polis & K. O. Winemiller (eds.), Food Webs: Integration of Patterns and Dynamics. Chapman & Hall, New York, NY: 81–95.

Vanni, M. J. & C. D. Layne, 1997. Nutrient recycling and herbivory as mechanisms in the 'top-down' effect of fish on algae in lakes. Ecology 78: 21–40.

Vanni, M. J., C. D. Layne & S. E. Arnott, 1997. 'Top-down' trophic interactions in lakes: effects of fish on nutrient cycling. Ecology 78: 1–20.

Vollenweider, A., 1976. Advances in defining critical loading levels for phosphorus in lake eutrophication. Mer. Ist. Ital. Idrobiol. 33: 53–83.

Wetzel, R. G. & R. A. Hough, 1973. Productivity and role of aquatic macrophytes in lakes: an assessment. Pol. Arch. Hydrobiol. 20: 9–19.

Hydrobiologia **404:** 145–156, 1999.
L.-A. Hansson & E. Bergman (eds), Nutrient Reduction and Biomanipulation as Tools to Improve Water Quality: The Lake Ringsjön Story.

Synthesis of theoretical and empirical experiences from nutrient and cyprinid reductions in Lake Ringsjön

E. Bergman[1], L.-A. Hansson[1], A. Persson[1], J. Strand[1], P. Romare[1], M. Enell[4], W. Granéli[1], J. M. Svensson[1], S. F. Hamrin[2], G. Cronberg[1], G. Andersson[3] & E. Bergstrand[2]
[1]*Department of Limnology, Insitute of Ecology, Ecology Building, University of Lund, SE-223 62 Lund, Sweden*
Present address: Department of Biology, Karlstad University, S-651 88 Karlstad, Sweden
[2]*Institute of Freshwater Research, SE-178 93 Drottningholm, Sweden*
[3]*County Administration Board of Scania, SE-205 15 Malmö, Sweden*
[4]*ITT Flygt AB, P.O. Box 1309, SE-171 25 Solna, Sweden*

Key words: biomanipulation, nutrient reduction, zooplankton, phytoplankton, bottom-up, top-down, phosphorus, submerged vegetation, benthic macrofauna

Abstract

The reduction in external phosphorus load to Lake Ringsjön during the 1980s, did not result in improved water transparency during the following ten-year period. Furthermore, a fish-kill in the Eastern Basin of the lake, in addition to a cyprinid reduction programme (biomanipulation; 1988–1992), in contrast to theory, did not lead to any increase in zooplankton biomass or size. This absence of response in the pelagic food chain may have been attributed to the increase in abundance of YOY (0+) fish, following the fish reduction programme. Despite the lack of effect on zooplankton, there was a decrease in phytoplankton biomass, a change in species composition and an increase in water transparency following biomanipulation. In 1989, one year after the fish-kill in Eastern Basin, the Secchi depth (summer mean) increased from 60 cm to 110 cm. In the following years, water transparency increased further, despite an increase in phosphorus loading. An unexpected effect of the biomanipulation was an increase in benthic invertebrate and staging waterfowl abundances, which occurred 2–4 years after fish reduction. Hence, the response in the benthic community following biomanipulation was considerably stronger than in the pelagic community. A likely explanation is that reduction in abundance of the benthic feeding fish species bream (*Abramis brama*), strongly affected the benthic invertebrate fauna. In this paper, we present what we believe happened in Lake Ringsjön, and which processes are likely to have been important at various stages of the restoration process.

Introduction

By providing drinking water, fishing opportunities and recreation, Lake Ringsjön has been of considerable value for the regional economy. It has, therefore, been subject to several attempts at improving water quality. In this chapter, we evaluate the attempts to improve the value of the lake by reducing the nutrient load and the abundance of cyprinid fish (biomanipulation). The fact that both of these large-scale restoration attempts were carried out, makes evaluation difficult as to which of them actually caused observed changes, since synergistic effects and lag-phases are likely to have occurred. Although knowledge of the mechanisms involved may not be of interest for lake managers, it is of considerable scientific interest to disentangle effects caused by the different measures. Expectations regarding the nutrient reduction programme were that reduced external load would lead to lower algal biomass and thereby clearer water. The nutrient reduction programme started in the late 1970s, yet by the end of the 1980s the water transparency had not noticeably improved, despite phosphorus loading being decreased from around 30 to 10 tons yr^{-1} (Figure 1). As a consequence of this, a biomanipulation programme was initiated.

The planning of biomanipulation in Lake Ringsjön was, as with other fish reductions at that time, mainly based on current food-chain theory (Hairston et al., 1960; Shapiro et al., 1975; Carpenter et al.,

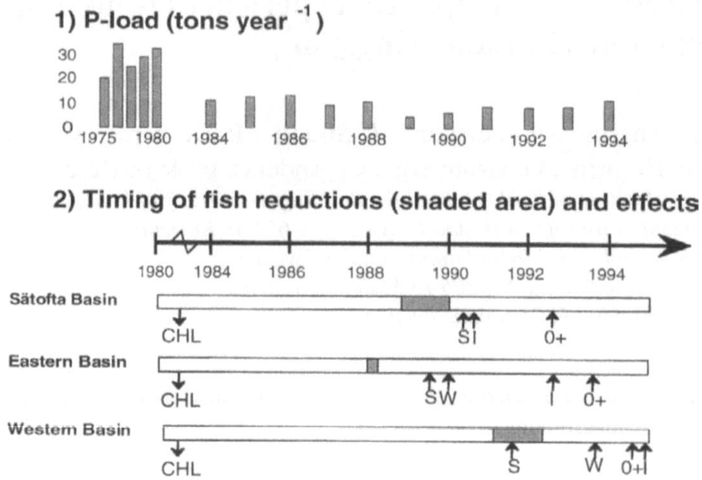

Figure 1. Time schedule behind the different remedial measures: (1) Phosphorus loading, and (2) Timing of the fish reductions and first recorded responses in several variables in the three basins. The response variables are: chlorophyll *a* (CHL); Secchi depth (S), abundance of 0+ fish (0+), abundance of benthic macroinvertebrates (I) and number of staging waterfowl (W) in the fall. Waterfowl were not counted in the Sätofta Basin.

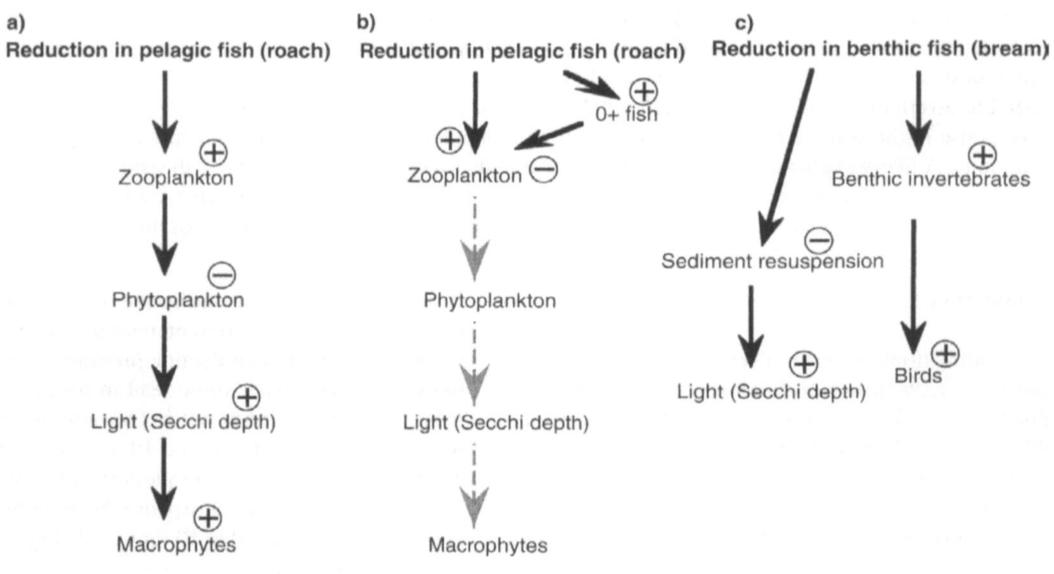

Figure 2. Possible mechanisms behind the changes in Lake Ringsjön following fish reduction. A (+) sign indicates an increase, and a (–) sign a decrease in biomass. (a) expected change in the pelagic food chain based on the food chain theory, (b) observed changes in the pelgaic food chain, illustrating the 'short-circuit' of the food chain effect due to 0+ fish expansion (the hatched arrows indicate processes of minor importance), (c) observed changes and suggested mechanisms in the benthic food web.

1985, 1987; Persson et al., 1988). It was, therefore, expected that removal of planktivorous fish should lead to considerable alterations in the pelagic food chain, including an increase in zooplankton abundance and mean size (Brooks & Dodson, 1965). As a result of higher grazing pressure imposed by an expanding zooplankton community, the phytoplankton biomass was expected to decline (Shapiro et al., 1975, Andersson et al., 1978), and water transparency was consequently expected to increase (Figure 2a). A recolonization of submerged macrophytes was also expected, which would contribute to preserving the lake in a clear-water state by a number of feedback mechanisms (Scheffer et al., 1993). Reality, however, differed somewhat from expectations. As expected, an increase in water transparency, and a decrease in phytoplankton biomass, occurred immediately following fish reduction (Bergman, 1999; Cronberg, 1999). However, the zooplankton community did not increase in biomass, the mean size of the larger grazer species had decreased, and not increased as expected (Bergman & Bergstrand, 1999) despite the fact that fish biomass decreased considerably (Hamrin, 1999; Bergman et al., 1999).

In this chapter, we aim at twinning the previous chapters together into a synthesis of what we believe happened in Lake Ringsjön, and which processes are likely to have been responsible for the observed changes. We start by describing the timing of improvements in water quality. Then we discuss changes in the phytoplankton community, the role of 0+ fish and their relation to the lack of a top-down response at the zooplankton level. We then discuss the responses observed among benthic organisms and waterfowl, as well as the possible lack of response in submersed macrophytes. Thereafter, we discuss the resuspension of phosphorus from the sediments and present our hypothesis of an 'indirect bottom-up effect', i.e. the question: "Were the positive effects on water transparency and phytoplankton biomass only due to the decrease in phosphorus loading, or were other 'non top-down' mechanisms, related to fish reduction, operating as well?". Finally, we speculate on the impact of other possibly important processes, not thoroughly covered by our sampling programme.

Timing of improvements

One important aspect when separating effects caused by different restoration efforts, may be to assess *when* improvements in water quality are observed. As seen in Figure 1, the major reductions in chlorophyll occurred between 1980 and 1981, which corresponds to the considerable decrease in external phosphorus load to the lake. It is notable however, that the decrease in algal biomass (chlorophyll *a*), following the nutrient reduction programme (Cronberg, 1999), was not reflected in any increase in water transparency before the fish reductions had taken place (Figure 1). This may indicate that water transparency was not only affected by algal biomass, but also by species composition (see below) of the algae, and by particles resuspended by benthic feeding fish (Lammens, 1989). The abundance of benthic invertebrates began to increase between 1–5 years, and waterfowl within 2 years after the biomanipulations (Figure 1). It should be noted however, that sampling of benthic invertebrates was not performed every year, making an exact timing of improvements difficult to assess. The increase in YOY (0+) cyprinid fish occurred more than 5 years after the fish reductions in Eastern Basin, where the reduction of adult cyprinid fish was highest (Hamrin, 1999; Bergman et al., 1999). In the other basins, a considerable increase in the YOY abundance was observed 2–3 years after the biomanipulations (Figure 1).

Dramatic quantitative and qualitative changes in the phytoplankton assemblage

Around 1900, Lake Ringsjön was an oligotrophic lake dominated by diatoms and chrysophytes. As the result of increasing external nutrient input to the lake, a dramatic increase in algal biomass and frequent blooms of blue-green algae occurred between 1970 and 1980. After 1980, there was a considerable decrease in phosphorus loading from about 30 tons yr^{-1} in 1980 to 10 tons yr^{-1} in 1984 (Figure 1). Hence, the quantitative reduction in algae was most probably a result of the nutrient reduction programme. However, the qualitative composition of the algal assemblage did not change notably before the fish reduction programme was completed (Cronberg, 1999). Then, *Microcystis* was to a certain extent replaced by *Anabaena* and *Aphanizomenon*, and the biodiversity of algae increased. There was also a noticeable reduction in blue-green picoplankton and small monads (2–8 μm), and in 1989–1991 the large, grazer resistant dinoflagellate *Ceratium hirundinella* was recorded in high abundances (Cronberg, 1999). This shift towards larger algal species, which may indicate increased grazing pressure by zooplankters following fish reduction, was most probably a major factor in the improved water transparency (Figure 1).

Although it is difficult to distinguish between effects caused by the nutrient and fish reductions respectively, it may be postulated that the *quantitative* reduction of blue-green algae would most probably not have occurred, or would have been less pronounced, if the decrease in nutrient input to the lake had not taken place. On the other hand, the *qualitative* changes towards larger species and higher diversity is likely to have been caused mainly by fish reduction.

Increased fish reproduction and recruitment

A typical response in lakes where the abundance of adult fish has been reduced is a 'boom' in abundance of young fish (Jeppesen et al., 1990b; Horppila et al., 1998; Romare & Bergman, 1999). Lake Ringsjön was no exception, and the increase in abundance of YOY (0+) fish occurred 1–5 years after each fish reduction (Figure 1). Since all 0+ fish in Lake Ringsjön are zooplanktivorous, the predation pressure exerted on zooplankton after fish reduction was higher than, or at least equally as high as, before adult planktivores were reduced (Romare & Bergman, 1999). In late summer, 0+ fish are size-selective zooplanktivores (Lessmark, 1983; van Densen, 1985), and several enclosure studies have shown that there is a negative correlation between 0+ fish density and zooplankton size (Post & McQueen, 1987; Dettmers & Stein, 1992; Hülsmann & Mehner, 1997; Romare et al., 1999). In Lake Ringsjön, we found a negative relationship between 0+ abundance and the size of *Daphnia* (mean size of animals larger than the smallest egg bearing female), in late summer and fall (August–November; Bergman & Bergstrand, 1999). This is in accordance with findings from other lakes (Vijverberg & van Densen, 1984; Cryer et al., 1986). Size-selective predation by fish has been shown to affect life history traits of daphnids, e.g. large daphnids initiate reproduction at a smaller size and produce smaller offspring in the presence of fish (Vanni, 1987; Gliwicz & Pijanowska, 1989; Hülsmann & Mehner, 1997; Sakwinska, 1998). When fish are removed, density of large *Daphnia* increases, and the contribution of the largest females to the total number of egg-bearing females increases (Pijanowska & Prejs, 1997). Nutrients on the other hand, have been shown to have less effect on the size of egg-bearing females than predation (Vanni, 1987). Hence, the change in mean daphnid size, and especially in the size of adult daphnids, is an indication of a change in the predation pressure from fish. The reduction of adult planktivores in Lake Ringsjön was expected to reduce predation pressure on large daphnids and thereby lead to a zooplankton community consisting of large-sized herbivores. Since this change in size did not occur, we suggest that 0+ fish, due to their high abundance in late summer, stopped the expected increase in zooplankton abundance and size, thereby decoupling the pelagic trophic cascade. Hence, the initial food chain concept has to be revised with respect to Lake Ringsjön (Figure 2 b).

Increased benthic invertebrate abundance

Many cyprinid fish are benthivorous feeders, and in systems where cyprinids dominate, these may not only affect the pelagic prey animals, such as zooplankters, but also the benthic macrofauna. This seems to be what happened in Lake Ringsjön, where there was no obvious top-down effect from the fish reduction on the zooplankton level, while there was a pronounced effect on the benthic macroinvertebrates (Svensson et al., 1999). One possible process is that an increase in submerged macrophyte covering provided structure and protection for benthic invertebrates. With respect to Lake Ringsjön, this is, however, a less likely explanation since the increase in macrophyte cover was local and relatively modest (as below). Also, the increase in abundance of predation-sensitive benthic macroinvertebrates occurred in all basins, independent of the distribution of the submerged vegetation. Moreover, it was mainly invertebrate species, not necessarily associated with macrophytes, that increased in abundance, such as chironomids, oligochaetes and molluscs (Svensson et al., 1999). We therefore suggest that the mechanism behind the increase in benthic invertebrates was the reduction in benthivorous cyprinid fish, such as bream (*Abramis brama*; Figure 2c).

Increased waterfowl abundance

A common aim of lake and wetland restorations is to improve the recreational potential. One recreational aspect is creating or re-establishing a lake suitable for waterfowl. The considerable increase in staging waterfowl numbers was, therefore, a welcomed result of the biomanipulation (Andersson & Nilsson, 1999).

This improvement was most likely a result of the increase in invertebrate abundance. However, not only diving ducks, which mainly feed on invertebrates, increased in numbers, but also coot (*Fulica atra*) and swans (*Cygnus* sp.), which mainly feed on vegetation, such as submerged macrophytes (Andersson & Nilsson, 1999). Their increase may indicate that the production and biomass of submerged macrophytes actually increased, but was consumed by increasing waterfowl populations. This explanation is especially plausible considering the fact that the highest number of waterfowl occurred in September–October, whereas sampling of macrophytes was generally performed in October. Hence, due to the focus on pelagic processes in the monitoring programme (see below), we may not have recorded a possible increase in macrophyte biomass.

Failed macrophyte expansion?

The concept of alternative stable states, and its relevance to the development of submerged vegetation, has been increasingly appreciated in biomanipulation projects (van Dijk & van Donk, 1991; Scheffer et al., 1993; Jeppesen et al., 1997). Whether or not a lake stays in a clear-water state following biomanipulation seems to depend, to a large extent, on the development of submerged vegetation (Jeppesen et al., 1997). Submerged macrophytes have been claimed to affect the lake ecosystem by a number of mechanisms, such as nutrient cycling and stabilization of the sediment surface (Carpenter & Lodge, 1986; Scheffer et al., 1993).

The poor development of submerged macrophytes in Lake Ringsjön after biomanipulation is likely to be one reason for the discrepancy between expected and observed results (Strand, 1999). The absence of refugia, such as macrophytes, for zooplankton, and the increase in the abundance of 0+ fish led to a decrease in mean size and biomass of zooplankton.

It is possible that the observed increase in abundance of grazing waterfowl in Lake Ringsjön has prevented the development of submerged macrophytes (see above and Andersson & Nilsson, 1999). It may be noted that the submerged macrophyte community in Lake Ringsjön is dominated by *Potamogeton* species, which are preferred over e.g. *Chara* and *Myriophyllum* as food by waterfowl (Weisner et al., 1997). Other studies have shown that waterfowl grazing can prevent, or at least delay, re-establishment of submerged macrophytes after a biomanipulation (e.g. Lauridsen et al., 1993). However, the most likely explanation for the poor development of submerged macrophytes in Lake Ringsjön is the lack of light reaching the areas suitable for colonization. Apparently, the increase in water transparency was not high enough for the establishment of submerged macrophytes (Strand, 1999). One reason for this could be the morphometry of the lake. It has been stated that a lake should have large enough (30% of the lake area) shallow areas (< 2 m) in order for positive effects of submerged macrophytes to influence the whole ecosystem (Hosper & Meijer, 1992). None of the basins in Lake Ringsjön has this high percentage of shallow areas, with the Sätofta Basin coming the closest with 27% lower than 2 m water depth, whereas Eastern and Western Basins have 11 and 16% (Strand, 1999). It is, therefore, possible that Lake Ringsjön is less than optimal for successful biomanipulation.

Reduced resuspension of sediment particles by benthic fish

Cyprinid fish can affect the amount of resuspended sediment and detritus, which could directly link the abundance of cyprinids and water quality (Andersson et al., 1978; Horppila & Kairesalo, 1990; Meijer et al., 1990; Brabrand et al., 1990). In Lake Ringsjön, a decrease in sediment resuspension, due to the decrease in biomass of benthivorous fish, may have affected light penetration into the water-column, and thus increased water transparency (Figure 2 c). One indication that benthic fish affected the resuspension of sediments, is that water transparency did not increase until after fish reduction, despite earlier decrease in algal biomass (Figure 1). Unfortunately, we have no measurements of sediment resuspension and its possible effect on phosphorus leaking. We do know though, that the lake has shifted from mainly being a phosphorus source to being more of a phosphorus sink (Granéli, 1999). Whether this shift is due to changes in pH, redox potential, bioturbation or an alteration in the sediment–water equilibria, due to the drastic decrease in external input, is largely unknown.

The indirect bottom-up effect

The recycling of nutrients by cyprinid fish could be the important link between the abundance of cyprinids

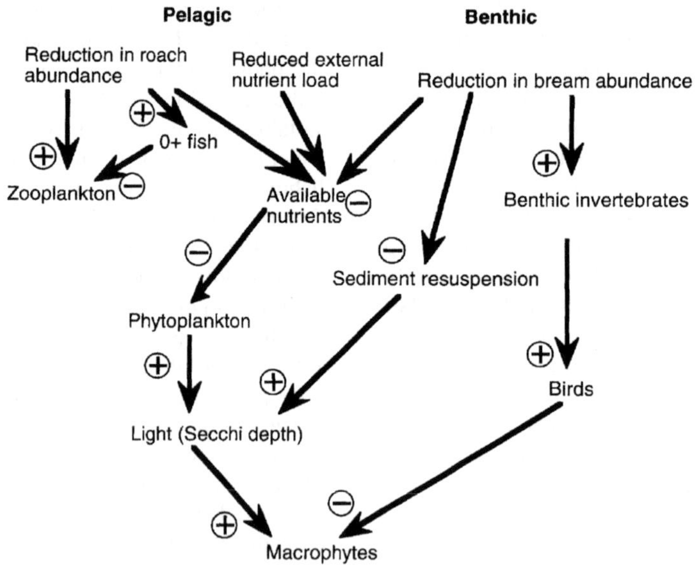

Figure 3. Suggested processes behind observed changes in different trophic levels in pelagial and littoral zones of Lake Ringsjön. A (+) sign indicates an increase, and a (–) sign a decrease. Starting at the top, the figure shows pelagic processes (reductions in external phosphorus load, and abundance of roach (*Rutilus rutilus*)), and benthic processes (reduction in abundance of bream (*Abramis brama*)). The pelagic food chain effect (roach) is short-circuited by the expansion of 0+ fish leading to no response in zooplankton. The reduction in benthic feeding fish (bream), however, leads to an increase in benthic invertebrates and a reduction in sediment resuspension, the latter improving the light penetration (water transparency). Increased abundance of benthic invertebrates attracted staging waterfowl which in turn affected macrophytes negatively.

and nutrient availability for primary producers (Figure 3) (Andersson et al., 1978; Horppila & Kairesalo, 1990; Meijer et al., 1990; Brabrand et al., 1990). The availability of nutrients not only affects the primary production, but also indirectly affects trophic levels higher than the primary producers. Therefore, this link could be the mechanism behind what we have called an 'indirect bottom-up effect', i. e. the cyprinid fish directly affects nutrient availability, which in turn affects all the other trophic levels. In Lake Ringsjön, this is supported by the decrease in phosphorus concentration and phytoplankton biomass, following the fish reduction and the lack of a relationship between zooplankton abundance and phytoplankton biomass (Bergman, 1999).

In 1998, Horppila et al. suggested that cyprinids 'take over' the top-down cascade and directly affect nutrient availability in the water by excreting nutrients, by bioturbation at the sediment surface and by changes in their abundance (e.g. mortality and survival). This means that a reduction in the abundance of benthivorous fish could improve water quality without directly involving the zoo- and phytoplankton levels. This is in accordance with our results in Lake Ringsjön. We believe however, that the effect of fish mortality, and thus the phosphorus content of fish, is of secondary importance. The reason for this view is firstly, that the extensive fish-kill in the Eastern Basin (all the dead fish were left in the lake) produced effects resembling those observed in the two basins where the fish were selectively removed from the lake. Secondly, a total of 200–250 tons of fish was removed from the Sätofta and Western Basins, which is equivalent to about 1 ton of phosphorus (assuming that 0.5% of the wet mass of a fish is phosphorus; Nakashima & Legget, 1980; Schindler et al., 1993). This represents only 10% of the annual phosphorus loading, and thus is not likely to be the primary mechanism behind the decrease in phytoplankton biomass and increase in water transparency (but see Sas, 1989; Jeppesen et al., 1990b). As the fish biomass in Lake Ringsjön has remained low since the fish reduction, it is unlikely that large amounts of phosphorus are being absorbed by, and retained in, the new fish community. The large amount of 0+ fish that grow each year act more as a nutrient recycler. The phosphorus that they acquire during the summer should be released during winter when mortality is high and, thus, be available for phytoplankton in the spring. Hence, we believe that reduced external phosphorus loading, fish community metabolism and bioturbation were more important for the decreasing nutrient level in the lake water than the phosphorus content of removed or dead fish.

Fish reduction efficiency

During the monitoring programme following the different remedial measures in Lake Ringsjön, we have compared the moderate improvements in water quality and the lack of an obvious pelagic top-down effect in Lake Ringsjön with the results of biomanipulations in other lakes. From this comparison, several questions about the fish reduction arose. One question is how large and how fast the reduction of planktivorous fish biomass has to be in order to succeed? Were the reductions too small and too slow in Lake Ringsjön? The aim of the biomanipulation in Lake Ringsjön was to reduce planktivore biomass by 80%. In retrospect, our estimations show that only the fish-kill in Eastern Basin resulted in such a high reduction. Possibly due to fish migrations between basins, only 60% and 50% of the planktivore biomass was removed in the Sätofta and Western Basins, respectively. Furthermore, the total time required for the fish reductions, including the fish-kill in Eastern Basin, was 5 years. Relative to some other biomanipulated lakes in Scandinavia (Lake Finjasjön, Lake Vesijärvi, Lake Vaeng; Hansson et al., 1998), this is a low reduction over a long period of time, which may explain the development of high abundances of 0+ fish afterwards (Romare & Bergman, 1999). Hence, the rate of growth and reproduction of cyprinids, to a large extent, compensated for the biomass taken out from the lake.

Another important aspect is the relative proportion of planktivores and piscivores. Jeppesen et al. (1990a, b) have shown in a large number of Danish lakes that the percentage of piscivores should be at least 50% of total fish biomass. In Lake Ringsjön, we had at most 40% piscivores in Eastern Basin, but only 10% in the Sätofta and 14% in Western Basins. The relatively high biomass of adult planktivores (although at a lower biomass than before the fish reductions), in combination with high 0+ abundances, have probably prevented the pelagic top-down effect we expected after the fish reduction. In some lakes, such as Lake Finjasjön (Annadotter et al., 1998) and Lake Vesijärvi (Horppila et al., 1998), reduction of planktivores has been complemented with a stocking of young piscivores. We believe that the top-down effect on the zooplankton level would have been more powerful if:

(a) The high 0+ abundance was lowered by selectively capturing and removing 0+ fish, and
(b) Piscivores were stocked to increase predation pressure on 0+ fish.

Other possible mechanisms: suggestions for future research

The Lake Ringsjön study clearly illustrates that during the period when the biomanipulation was performed, the simple food-chain theory dominated completely as background concept (see Bergman et al., 1999 and references therein), neglecting littoral and benthic variables. During the analysis of data, we learned that other, mainly benthic, variables responded even more than the pelagic ones. Unfortunately, most of these variables were not recorded as thoroughly as the pelagic variables, and we therefore have less reliable data on most of them. Moreover, some of them were not recorded at all, and we can only speculate as to their possible impact and suggest that they are included in future research! Below we discuss some of the more important of these variables.

Periphytic algae

One possible variable that might aid in maintaining the lower phosphorus concentration in the water after an improved light climate following cyprinid reduction, is periphytic algae growing at the sediment surface. Periphytic algae could influence phosphorus recirculation in the lake by absorbing phosphorus from the water, and by covering a large area of the lake sediment as a 'lid', thereby reducing sediment resuspension. This algal 'lid' also increases the oxygen content of the sediment which reduces phosphorus release by sediments (Boström et al., 1982). The distribution of periphytic algae may have been favored by the increased water transparency and reduced bioturbation from bream. We have no data or observations on this variable, although several studies suggest that periphytic algae considerably affect the phosphorus dynamics at the sediment–water interface (Carlton & Wetzel, 1988; Hansson, 1990).

Furthermore, submerged macrophytes, which rely mostly on sediment interstitial water for nutrients

(Carignan & Kalff, 1980), provide a substrate on which periphytic algae can grow, enabling them to compete directly with phytoplankton. Periphyton production in Lake Ringsjön was estimated to 0.1–0.5 μg chl *a* cm^{-2} d^{-1} (0.1 g AFDW m^{-2} d^{-1}), and the biomass of periphytic algae per gram AFDW macrophyte (*Potamogeton pectinatus*) was estimated to 0.1–0.4 g AFDW depending on wind exposure (Strand & Weisner, 1996). It is hard to say how strong these mechanisms are in Lake Ringsjön since no biomass measurements of submerged macrophytes have been performed. It is possible to roughly estimate the potential amount of phosphorus incorporated into periphytic algae by assuming 100 g DW of submerged macrophytes to 1.5 m water depth in Sätofta and Western Basins and to 1 m in Eastern Basin (giving about 10% of the lake area covered = 4 km^2). The surface area of *Potamogeton pectinatus*, which is one of the most common macrophytes in the lake (Strand, 1999), is 0.16 m^2 g^{-1} DW (Eriksson & Weisner, 1997). Periphyton phosphorus content varies between 0.4 and 2.4% of the dry weight (Davis et al., 1990), and if we use the rather conservative value of 1% phosphorus per g periphyton DW and the above-mentioned outer depth of submerged macrophytes, we get an uptake of 0.064 tons of phosphorus d^{-1} in the lake. This means that during summer (100 days), about 6 tons of phosphorus is removed from the water column and locked into periphyton biomass. Hence, it is not unlikely that periphyton on macrophytes may have contributed to the low phosphorous concentrations in the water. These calculations however, are by no means exact, and the estimate should therefore be viewed as a thought-experiment about a potentially important variable to include in future studies.

Littoral vs. *pelagic processes*

As stated above, the expected top-down effect in the pelagial of Lake Ringsjön may have been short-circuited by the expansion of 0+ fish. However, in the littoral we found an increased biomass of predation-sensitive macroinvertebrates, suggesting a littoral top-down effect of the fish reduction (Figures 2 & 3). This leads to the question: how important is the littoral zone in relation to the pelagic zone? To answer this question, we need to know more about the distribution and density of macrophytes and fish migrations between pelagial and littoral habitats. Macrophytes create a refuge from fish predation for 0+ cyprinids and zooplankton (Schriever et al., 1995; Stansfield et al., 1997), but also favour the occurrence of pike (*Esox lucius*), that could feed on 0+ fish (Benndorf, 1990; Grimm & Backx, 1990). In Lake Ringsjön, macrophyte stands were not very dense and there was no major change in the stands over the observed years (Strand, 1999), so we do not expect vegetation in the littoral to have an important role. The role of fish migrations on nutrient translocation is a topic currently being investigated by others (Schindler et al., 1993, 1996; Deegan, 1993). Although we have not measured fish migration or nutrient excretion by fish in Lake Ringsjön, the littoral is most probably closely linked to the rest of the lake ecosystem as indicated by the considerable effect on benthic macroinvertebrates following fish reduction (Svensson et al., 1999).

Benthic processes

Traditionally, exchange processes between sediment and water have been studied in microcosms, not including the effects of benthic fish activities. Such experiments have shown the importance of benthic invertebrates for the exchange of nutrients between sediment and water (Gallepp et al., 1978; Gallepp, 1979; Granéli, 1979; Starkel, 1985), as well as the physical and chemical characteristics of the sediment itself (reviewed in Boström et al., 1982). The fish reduction program of Lake Ringsjön clearly affected the benthic food chain more than the pelagic one. However, the processes actually causing the observed lowering in nutrient concentrations are not fully understood, although there are many suggestions both in this volume and in the literature as a whole (see e.g. Søndergaard et al., 1990). Resuspension of sediment material by benthic feeding bream has, in the shallow Dutch lakes, been shown to be of large importance for both the water transparency and the nutrient concentration in the water (Meijer et al., 1990). However, Lake Ringsjön is deeper than the Dutch lakes, and resuspension should consequently be of less importance given its lower surface to volume ratio. There might be other processes, linked to the foraging of benthic fish, that have been overlooked. Perhaps trophic cascades are just as important for benthic community dynamics, as it has been shown to be in littoral (Brönmark et al., 1992), and pelagic communities (Carpenter et al., 1985, 1987; Vanni et al., 1990). Indeed, the most pronounced effects were observed for the benthic invertebrate community (Svensson et al., 1999). This is a line of experimental research that up to now has been neglected, maybe because of the logistic difficulties introduced by combining soft sediments and large fish. It should be noted that most lakes in the world are small, with a high surface to volume ratio. Consequently, a large part of the fish biomass in such lakes may be built up via benthic producers and detrivores. It might therefore be of importance to consider how the abundance and behaviour of various benthic populations is affected by top-down forces, and in turn how that affects the exchange processes between sediment and water. Such a line of research could in the future provide some missing pieces in the puzzle of understanding lake ecosystem dynamics.

Synthesis

The reduction of the external nutrient input to Lake Ringsjön reduced the concentration of phosphorus in the lake considerably. Despite the minor improvements in water transparency, this reduction in external input was most probably a prerequisite for improvements following the fish reduction programme (Jeppesen et al., 1990a,b; Bergman et al., 1999). Most probably, the water quality would eventually have improved even if no other measures than nutrient reduction had been carried out, although the time scale may be inappropriate from a lake management perspective (Søndergaard et al., 1990).

With respect to the biomanipulation (including the fish-kill in Eastern Basin), our expectation was that a reduction in cyprinid fish abundance would lead to increased abundance and size of zooplankton and thereby to a reduced algal biomass (Figure 2a). At the time when the biomanipulation was planned in Lake Ringsjön, this food chain effect was believed to be *the only* major mechanism involved. One of the major lessons learned from the Lake Ringsjön biomanipulation is that this is not the whole truth. Instead, our results point out that fish removal 'triggers' several other processes. One is the tremendous increase in recruitment of 0+ fish, which may 'short-circuit' the food-chain effct by imposing a similar predation pressure on zooplankton as the adult fish did before biomanipulation (Figure 2b). Hence, this short-circuit is the likely explanation of the lack of an increase in zooplankton abundance and size in Lake Ringsjön.

Another mechanism involved is high-lighted by the increase in abundance of benthic invertebrates. A likely explanation of this is reduced predation pressure from benthic feeding fish as a result of the biomanipulations (Figure 2c). In addition, reduced populations of benthic feeding fish should decrease the amount of resuspended sediment particles, thereby reducing the availability of sediment-bound nutrients to organisms in the water column; a mechanism most likely involved in the improved Secchi-depth readings following biomanipulations (Figures 1 & 2c). The improved light climate might also have improved conditions for sediment dwelling algae (periphytic algae), constituting a main food source for benthic invertebrates. Hence, the removal of benthic feeding fish may have improved the conditions for benthic invertebrates, both by increasing availability of food and reducing predation.

The removal of benthic feeding fish appears to have considerable effects, not only on the light climate and the abundance of benthic invertebrates, but also on the abundance of macrophytes, as has been shown in previous studies (Jeppesen et al., 1997). In Lake Ringsjön however, the submerged macrophytes did not show any noticeable increase in biomass or cover, which was surprising since large areas of Sätofta and Western Basins were covered by macrophytes before the eutrophication process began (Strand, 1999). One possible explanation to the lack of a response in macrophytes may be the increased abundances of staging waterfowl, especially swans and coots, which feed on macrophytes (Lauridsen et al., 1993; Weisner et al., 1997; Andersson & Nilsson, 1999). However, the improvement in light climate was relatively modest, and may not have been enough to trigger a major expansion of macrophytes (Strand, 1999). Since Lake Ringsjön is relatively large and deep, the fish removal should have been much more efficient than it was, in order to reach desirable conditions for macrophytes (Figure 6.2 in Scheffer, 1998).

Our conclusion regarding the processes involved in the Lake Ringsjön biomanipulation is summarized in Figure 3, showing reductions in abundances of roach (*Rutilus rutilus*) and bream (*Abramis brama*) at the top of the Figure. The expected positive effect on zooplankton abundance and size, due to reduction in roach abundance, was short-circuited by the expansion of 0+ fish, leading to no response in zooplankton (see also Figure 2a). The reduction in benthic feeding fish (bream), however, led to an increase in benthic invertebrates and a reduction in sediment resuspension, the latter improving light penetration (water transparency; Figure 3). Increased abundance of benthic invertebrates attracted staging waterfowl. Many of these feed on macrophytes, which may have been one of the processes affecting macrophyte expansion negatively. Fish reduction also affected the availability of nutrients negatively by the 'indirect bottom-up effect', that is, reduced bioturbation and excretion of easily accessible phosphorus (phosphate). Hence, this mainly benthic explanatory model differs considerably from the theoretical concept in planning the biomanipulation (Figure 2a).

Although the restoration of Lake Ringsjön may be viewed as only moderately successful for the lake itself, our knowledge of the mechanisms and processes involved in a fish reduction programme has increased considerably. It is our hope that lessons learned from the Lake Ringsjön restoration attempts can serve as a basis for future attempts to improve water quality of eutrophicated lakes, and that our mistakes can be avoided.

References

Andersson, A., H. Berggren & G. Cronberg, 1978. Effects of planktivorous and benthivorous fish on organisms and water chemistry in eutrophic lakes. Hydrobiologia 59: 9–15.

Andersson, G. & L. Nilsson, 1999. Autumn waterfowl abundances in Lake Ringsjön, 1968–1996. Hydrobiologia 404: 41–51.

Annadotter, H., G. Cronberg, R. Ågren, L. Jönsson, B. Lundstedt, P.-Å. Nilsson & S. Ströbäck, 1998. Multiple techniques for the restoration of a hypereutrophic lake. In D. Harper, W. Brieley, G. Phillips & A. Ferguson (eds), The Ecological Basis for Lake and Reservoir Management. Wiley, Chichester.

Benndorf, J., 1990. Conditions for effective biomanipulation; conclusions derived from whole-lake exeriments in Europe. Hydrobiologia 200/201: 187–203.

Bergman, E. & E. Bergstrand, 1999. Lack of a top-down effect on the zooplankton community after a cyprinid reduction. Hydrobiologia 404: 77–87.

Bergman, E., 1999. Changes in nutrient load and lake water chemistry in Lake Ringsjön, southern Sweden, from 1966 to 1996. Hydrobiologia 404: 9–18.

Bergman, E., L.-A. Hansson & G. Andersson, 1999. Biomanipulation in a theoretical and historical perspective. Hydrobiologia 404: 53–58.

Boström, B., M. Jansson & C. Forsberg, 1982. Phosphorus release from lake sediments. Arch. Hydrobiol. Ergebn. Limnol. 18: 5–59.

Brabrand, Å., B. A. Faafeng & J. P. M. Nilssen, 1990. Relative importance of phosphorus supply to phytoplankton production: Fish excretion versus external loading. Can. J. Fish Aquat. Sci. 47: 364–372.

Brooks, J. L. & S. I. Dodson, 1965. Predation, body size and composition of plankton. Science 150: 28–35.

Brönmark, C., S. P. Klosiewski & R. A. Stein, 1992. Indirect effects of predation in a freshwater, benthic food chain. Ecology 73: 1662–1674.

Carignan, R. & J. Kalff, 1980. Phosphorus sources for aquatic weeds: water or sediments? Science 207: 987–988.

Carlton, R. & R. Wetzel, 1988. Phosphorus flux from lake sediments: effects of epipelic algal oxygen production. Limnol. Oceanogr. 33: 562–570.

Carpenter, S. R. & D. M. Lodge, 1986. Effects of submersed macrophytes on ecosystem processes. Aquat. Bot. 26: 341–370.

Carpenter, S. R., J. F. Kitchell & J. R. Hodgson, 1985. Cascading trophic interactions and lake productivity. BioScience 35: 634–639.

Carpenter, S. R., J. F. Kitchell, J. R. Hodgson, P. A. Cochran, J. J. Elser, M. M. Elser, D. M. Lodge, D. Kretchmer, X. He & C. N. von Ende, 1987. Regulation of lake primary production by food web structure. Ecology 68: 1863–1878.

Cronberg, G., 1999. Qualitative and quantitative investigations of phytoplankton in Lake Ringsjön, Scania, Sweden. Hydrobiologia 404: 27–40.

Cryer, M., G. Peirson & C.R. Townsend, 1986. Reciprocal interactions between roach, *Rutilus rutilus*, and zooplankton in a small lake: prey dynamics and fish growth and recruitment . Limnol. Oceanogr. 31: 1022–1038.

Davis, L. S., J. P. Hoffman & P. W. Cook, 1990. Production and nutrient accumulation by periphyton in a wastewater treatment facility. J. Phycol. 26: 617–624.

Deegan, L. A., 1993. Nutrient and energy transport between estuaries and coastal marine ecosystems by fish migration. Can. J. Fish. Aquat. Sci. 50: 74–79.

Dettmers, J. M. & R. A. Stein, 1992. Food consumption by larval gizzard shad: zooplankton effects and implications for reservoir communities. Trans. Am. Fish. Soc. 121: 494–507.

Eriksson, P. G. & S. E. B. Weisner, 1997. Nitrogen removal in a wastewater reservoir: the importance of denitrification by epiphytic biofilms on submersed vegetation. J. Envir. Qual. 26: 905–910.

Gallepp, G. W., 1979. Chironomid influence on phosphorus release in sediment-water microcosms. Ecology 60: 547–556.

Gallepp, G. W., J. F. Kitchell & S. M. Bartell, 1978. Phosphorus release from lake sediments as affected by chironomids. Verh. Int. Ver. Limnol. 20: 458–465.

Gliwicz, Z. M. & J. Pijanowska, 1989. Succession in plankton communities. In U. Sommer (ed.), Plankton Ecology. Springer-Verlag, Berlin: 253–296.

Granéli, W., 1979. The influence of *Chironomus plumosus* larvae on the exchange of dissolved substances between sediment and water. Hydrobiologia 66: 149–159.

Granéli, W., 1999. Internal phosphorus loading in Lake Ringsjön - the role of the sediments. Hydrobiologia 404: 19–26.

Grimm, M. P. & J. J. G. M. Backx, 1990. The restoration of shallow eutrophic lakes, and the role of northern pike, aquatic vegetation and nutrient concentration. Hydrobiologia 200/201: 557–566.

Hairston, N. G., F. E. Smith & L. B. Slobodkin, 1960. Community structure, population control and competition. Am. Nat. 94: 421–425.

Hamrin, S. F., 1999. Planning and execution of the fish reduction in Lake Ringsjön. Hydrobiologia 404: 59–63.

Hansson, L.-A., 1990. Quantifying the impact of periphytic algae on nutrient availability of phytoplankton. Freshwat. Biol. 24: 265–273.

Hansson, L.-A., H. Annadotter, E. Bergman, S. F. Hamrin, E. Jeppesen, T. Kairesalo, E. Luokkanen, P.-E. Nilsson & M. Søndergaard, 1998. Biomanipulation as an application of food chain theory: constraints, synthesis and recommendations for temperate lakes. Ecosystems 1: 558–574.

Horppila, J. & T. Kairesalo, 1990. A fading recovery: the role of roach (*Rutilus rutilus* L.) in maintaining high phytoplankton productivity and biomass in Lake Vesijärvi, southern Finland. Hydrobiologia 200/201: 153–165.

Horppila, J., H. Peltonen, T. Malinen, E. Luokkanen & T. Kairesalo, 1998. Top-down or bottom-up effects by fish - issues of concern in biomanipulation of lakes. Restoration Ecology 6: 1–10.

Hosper, H. & M.-L. Meijer, 1992. Biomanipulation, will it work for your lake? Simple test for the assessment of chances for clear water following drastic fish-stock reduction in shallow, eutrophic lakes. INTECOL's 4th International Wetlands Conference, Columbus, Ohio, U.S.A.

Hülsmann, S. & T. Mehner, 1997. Predation by underyearling perch (*Perca fluviatilis*) on a *Daphnia galeata* population in a short-term enclosure experiment. Freshwat. Biol. 38: 209–219.

Jeppesen, E., M. Søndergaard, E. Mortensen, P. Kristensen, B. Riemann, H. J. Jensen, J. P. Müller, O. Sortkjær, J. P. Jensen, K. Christoffersen, S. Bosselmann & E. Dall, 1990a. Fish manipulation as a lake restoration tool in shallow, eutrophic temperate lakes 1: cross-analysis of three Danish case-studies. Hydrobiologia 200/201: 205–218.

Jeppesen, E., J. P. Jensen, P. Kristensen, M. Søndergaard, E. Mortensen, O. Sortkjær & K. Olrik, 1990b. Fish manipulation as a lake restoration tool in shallow, eutrophic temperate lakes 2: threshold levels, long-term stability and conclusions. Hydrobiologia 200/201: 219–227.

Jeppesen, E., J. P. Jensen, M. Søndergaard, T. Lauridsen, L. J. Pedersen & L. Jensen, 1997. Top-down control in freshwater lakes: the role of nutrient state, submerged macrophytes and depth. Hydrobiologia 342/343: 151–164.

Lammens, E. H., 1989. Causes and consequences of the success of bream in dutch eutrophic lakes. Hydrobiol. Bull. 23: 11–18.

Lauridsen, T. L., E. Jeppesen & F. Østergaard Andersen, 1993. Colonization of submerged macrophytes in shallow fish manipulated Lake Væng: impact of sediment composition and waterfowl grazing. Aquat. Bot. 46: 1–15.

Lessmark, O., 1983. Competition between perch (*Perca fluviatilis*) and roach (*Rutilus rutilus*) in south Swedish lakes, Ph. D. thesis, Univ. of Lund, Lund, Sweden.

Meijer, M.-L., M. W. de Haan, A. W. Breukelaar & H. Buiteveld, 1990. Is reduction of the benthivorous fish an important cause of high transparancy following biomanipulation in shallow lakes? Hydrobiologia 200/201: 303–315.

Moss, B., J. Madgwick & G. Phillips, 1996. A guide to the restoration of nutrient-enriched shallow lakes. Broads Authority, U.K.

Nakashima, B. S. & W. C. Legget, 1980. Natural sources and requirements of phosphorus for fishes. Can. J. Fish. Aquat. Sci. 37: 679–686.

Persson, L., G. Andersson, S. F. Hamrin & L. Johansson, 1988. Predator regulation and primary production along the productivity gradient of temperate lake ecosystems. In S. R. Carpenter (ed.), Complex Interactions in Lake Communities. Springer-Verlag, New York: 45–65.

Pijanowska, J. & A. Prejs, 1997. Food-web manipulation in shallow, eutrophic lakes: bridging the gap between the whole-lake approach and behavioural and demographic studies. Hydrobiol. 342/343: 305–310.

Post, J. R. & D. J. McQueen, 1987. The impact of planktivorous fish on the structure of a plankton community. Freshwat. Biol. 17: 79–89.

Romare, P. & E. Bergman, 1999. Juvenile fish expansion following biomanipulation and the resulting effect on the predation pressure on zooplankton. Hydrobiologia 404: 89–97.

Romare, P., E. Bergman & L.–A. Hansson, 1999. The impact of larval and juvenile fish on zooplankton and algal dynamics. Limnol. Oceanogr. (In Press).

Sakwinska, O., 1998. Plasticity of *Daphnia magna* life history traits in response to temperature and information about a predator. Freshwat. Biol. 39: 681–687.

Sas, H., 1989. Lake restoration by reduction of nutrient loading. Expectation, experiences, extrapolation. Acad. Ver. Richarz GmbH, 497 pp.

Scheffer, M., 1998. Ecology of shallow lakes. Population and community biology series 22. Chapman and Hall, London.

Scheffer, M., S. H. Hosper, M.-L. Meijer, B. Moss & E. Jeppesen, 1993. Alternative Equilibria in shallow lakes. TREE 8: 275–279.

Schindler, D. E., J. F. Kitchell, X. He, S. R. Carpenter, J. R. Hodgson & K. L. Cottingham, 1993. Food web structure and phosphorus cycling in lakes. Trans. Am. Fish. Soc. 122: 756–772.

Schindler, D. E., S. R. Carpenter, K. L. Cottingham, X. He, J. R. Hodgson, J. F. Kitchell & P. A. Soranno, 1996. Food web structure and littoral zone coupling to pelagic trophic cascades. In G. A. Polis & K. O. Winemiller (eds), Food Webs. Integration of Patterns and Dynamics. Chapman and Hall, New York, N.Y.

Schriever, P., J. Bøgestrand, E. Jeppesen & M. Søndergaard, 1995. Impact of submersed macrophytes on fish-zooplankton-phytoplankton interactions: large-scale enclosure experiments in a shallow eutrophic lake. Freshwat. Biol. 33: 255–270.

Shapiro, J., V. Lamarra & M. Lynch, 1975. Biomanipulation: an ecosystem approach to lake restoration. In P. L. Brezonik & J. L. Fox (eds), Proceedings of the Symposium on Water Quality Management Through Biological Control, 85–96. Rep. No. ENV-07-75-1.

Stansfield, J. H., M. R. Perrow, L. D. Tench, A. J. D. Jowitt & A. A. L. Taylor, 1997. Submerged macrophytes as refuges for grazing Cladocera against fish predation: observations on seasonal changes in relation to macrophyte cover and predation pressure. Hydrobiologia 342/343: 229–240.

Starkel, W. M., 1985. Predicting the effects of macrobenthos on the sediment/water flux of metals and phosphorus. Can. J. Fish. Aquat. Sci. 42: 95–100.

Strand, J., 1999. The development of submerged macrophytes in Lake Ringsjön after biomanipulation. Hydrobiologia 404: 113–121.

Strand, J. A. & S. E. B. Weisner, 1996. Wave exposure related growth of epiphyton: implications for the distribution of submerged macrophytes in eutrophic lakes. Hydrobiologia 325: 113–119.

Svensson, J., E. Bergman & G. Andersson, 1999. Impact of cyprinid reduction on the benthic macroinvertebrate community and implications for increased nitrogen retention. Hydrobiologia 404: 99–112.

Søndergaard, M., E. Jeppesen, E. Mortensen, E. Dall, P. Kristensen & O. Sortkjær, 1990. Phytoplankton biomass reduction after planktivorous fish reduction in a shallow, eutrophic lake: a combined effect of reduced internal P-loading and increased zooplankton grazing. Hydrobiologia 200/201: 220–240.

van Densen, W. L. T., 1985. Feeding behaviour of major 0+ fish species in a shallow, eutrophic lake (Tjeukemeer, The Netherlands). Z. angew. Ichthyol. 2: 49–70.

van Dijk, G. M. & E. van Donk, 1991. Perspectives for submersed macrophytes in shallow lake restoration projects in the Netherlands. Hydrobiol. Bull. 24: 125–131.

Vanni, M. J., C. Luecke, J. F. Kitchell, Y. Allen, J. Temte & J. J. Magnusson. 1990. Effects on lower trophic levels of massive fish morality. Nature 344: 333–335.

Vanni, M. J., 1987. Effects of food availability and fish predation on a zooplankton community. Ecological Monographs 57: 61–88.

Vijverberg, J. & W. L. T. van Densen, 1984. The role of the fish in the foodweb of Tjeukemeer, The Netherlands. Verh. Int. Verein. Limnol. 22: 891–896.

Weisner, S. E. B., J. A. Strand & H. Sandsten, 1997. Mechanisms regulating abundance of submerged vegetation in shallow eutrophic lakes. Oecologia 109: 592–599.

MIX
Papier aus verantwortungsvollen Quellen
Paper from responsible sources
FSC® C105338

If you have any concerns about our products,
you can contact us on
ProductSafety@springernature.com

In case Publisher is established outside the EU,
the EU authorized representative is:
Springer Nature Customer Service Center GmbH
Europaplatz 3, 69115 Heidelberg, Germany

Printed by Libri Plureos GmbH
in Hamburg, Germany